Power Systems

Electrical power has been the technological foundation of industrial societies for many years. Although the systems designed to provide and apply electrical energy have reached a high degree of maturity, unforeseen problems are constantly encountered, necessitating the design of more efficient and reliable systems based on novel technologies. The book series Power Systems is aimed at providing detailed, accurate and sound technical information about these new developments in electrical power engineering. It includes topics on power generation, storage and transmission as well as electrical machines. The monographs and advanced textbooks in this series address researchers, lecturers, industrial engineers and senior students in electrical engineering.

Power Systems is indexed in Scopus

More information about this series at https://link.springer.com/bookseries/4622

Albana Ilo · Daniel-Leon Schultis

A Holistic Solution for Smart Grids based on LINK–Paradigm

Architecture, Energy Systems Integration, Volt/var Chain Process

 Springer

Albana Ilo
TU Wien
Vienna, Austria

Daniel-Leon Schultis
TU Wien
Vienna, Austria

ISSN 1612-1287   ISSN 1860-4676 (electronic)
Power Systems
ISBN 978-3-030-81532-5   ISBN 978-3-030-81530-1 (eBook)
https://doi.org/10.1007/978-3-030-81530-1

This Springer imprint is published by the registered company Springer Nature Switzerland AG
The registered company address is: Gewerbestrasse 11, 6330 Cham, Switzerland

Preface

The work on the holistic architecture of smart grids has its roots in the research project FENIX.[1] Its goal was to control the reactive power exchange between the transmission and distribution grid using distributed generators' capabilities connected in a medium-voltage grid. The Volt/Var control in the medium-voltage grid was realised in an open loop. ZUQDE[2] project was crucial for the emergence of the holistic technical model of smart grids. The Volt/Var control was implemented in a closed loop, indicating that the secondary control, based on state estimator results, is technically feasible in a medium-voltage grid from an industrial point of view. At that time, I was with Siemens AG, Austria. I had the opportunity to thoroughly understand the different applications of energy and distribution management systems and participate in their design and implementation.

My research aimed to find a rational model to understand and design the future's power supply systems. The change at TU Wien was decisive to deal with this subject academically. During the research project of SmaGPa,[3] the smart grid paradigm *LINK* and the associated holistic architecture were developed.

The journey to introduce the holistic view and develop the *LINK*–Paradigm was challenging. It was often very depressing to pursue a different perspective than the usual one. I thank Prof. Wolfgang Gawlik, Head of the power system department, and Prof. Manfred Schrödl, Head of the Institute for Energy Systems and Electric Drives, for creating the primary conditions of my work.

I thank Daniel-Leon Schultis, my ex-student and my closest collaborator, for his engagement and excellent work: it was our joint decision to write this book.

I am grateful to many colleagues in various utilities worldwide who, through their discussions, have enabled me to know the different processes required to operate the power systems. I would also like to thank all of my colleagues at Siemens for their cooperation in carding various applications down to the smallest detail that has helped me get a solid understanding of transmission and distribution management systems.

---

[1] FENIX project funded under FP6 by the European Commission.

[2] ZUQDE project funded by "Neue Energien 2020" of "Klima- und Energiefonds", Austria.

[3] SmaGPa project funded by TU Wien, Austria.

Finally, I am incredibly grateful to my mother, Iliana. As former Vice Director of the Publishing House of the University of Tirana, she has repeatedly encouraged me to write and finish this book. I cannot go here without mentioning my youngest son Thomas. Although still a very young man, he was a worthy debater of various analytical methods in physics and mathematics.

Vienna, Austria   Albana Ilo
June 2021

Contents

About the Authors

Albana Ilo studied electrical engineering at the University of Tirana, Albania. In 1991, she moved to Austria and later received her doctorate from TU Wien with the topic "Flux distribution and power loss in transformer cores as a function of joint design" in 1998. She switched then to industry (Siemens AG Austria), where she was working on many projects worldwide as Recognised Expert, and later as Principal Key Expert Consultant in power systems. In 2013, she moved back to TU Wien. In 2020, she obtained the "Venia Docendi", thus qualifying her to teach in the area of "Electrical Energy Systems" with a habilitation on "*LINK*-based holistic architecture for future power systems".

She works on integrating distributed generation or smart grids and their impact on transmission and distribution networks. Her merit is the *LINK*–Paradigm and the holistic solution for smart grids derived from it that enables the decarbonisation of the economy (Sector Coupling) and the democratisation of the electricity industry (Energy Communities).

Daniel-Leon Schultis received the B.Sc. degree in electrical engineering at the Graz University of Technology in 2016 and the Dipl.-Ing. (equivalent to M.Sc.) from TU Wien in 2017. Since 2017, he has been working as Project Assistant at the Institute of Energy Systems and Electrical Drives, TU Wien. In September 2021, he received his doctorate in Electrical Engineering at TU Wien with the theme "Vertical Volt/Var chain control as part of the *LINK*-based holistic architecture". His works are honoured with several awards, including "Austria's Energy Award 2019".

Abbreviations

AC	Alternating Current
AGC	Automatic Generation Control
APA	Active Power Appliance
BGA	Balancing Group Area
BLiN	Boundary Link Node
BPN	Boundary Producer Node
BRP	Balance Responsible Party
BSN	Boundary Storage Node
BVL	Boundary Voltage Limit
CAM	Control Area Manager
CC	Coupling Component
CD	Compensating Device
CHP	Combined Heat and Power
CoCe	Control Centre
CP	Customer Plant
CPG	Customer Plant Grid
CSA	Clearing and Settlement Agent
CVPP	Commercial Virtual Power Plant
CVR	Conservation Voltage Reduction
DC	Direct Current
DEG	Dynamic Equivalent Generator
DER	Distributed Energy Resources
Dev	Device
DFIG	Doubly Fed Induction Generator
DG	Distributed Generation
DiC	Direct Control
DMS	Distribution Management System
DR	Demand Response
DSO	Distribution System Operator
DSSE	Distribution System State Estimator
DTR	Distribution Transformer
EC	Energy Community

EI	Equivalent Impedance
ElA	Electrical Appliance
EMS	Energy Management System
EPO	Electricity Producer-Link Operator
ESI	Energy Systems Integration
EV	Electric Vehicle
FACTS	Flexible Alternating Current Transmission Systems
FCWG	Full-Converter Wind Generator
FENIX	Flexible Electricity Network to Integrate the eXpected "Energy Evolution"
GDPR	General Data Protection Regulation
GriLiO	Grid-Link Operator
HESS	Hydrogen Energy Storage System
HMU	House Management Unit
HV	High Voltage
HVDC	High-Voltage Direct Current
HVG	High-Voltage Grid
HVSO	High-Voltage System Grid-Link Operator
ICT	Information and Communication Technology
IoT	Internet of Things
ISO	Independent System Operator
LC	Local Control
LCC	Line-Commutated Converters
LFC	Load Frequency Control
LRM	Local Retail Market
LV	Low Voltage
LVG	Low-Voltage Grid
LVR	Line Voltage Regulator
LVSO	Low-Voltage System Grid-Link Operator
MC	Manual Control
MSC	Mechanically Switched Capacitor
MSR	Mechanically Switched Reactor
MV	Medium Voltage
MVG	Medium-Voltage Grid
MVSO	Medium-Voltage System Grid-Link Operator
OLTC	On-Load Tap Changer
P2Ch	Power-to-Chemicals
P2G	Power-to-Gas
P2H&C	Power-to-Heat and Cold
P2HESS	Power-to-Hydrogen Energy Storage System
P2T	Power-to-Thermal
P2X	Power-to-X
PC	Primary Control
PhST	Phase-Shifting Transformer
PV	Photovoltaic

RES	Renewable Energy Resources
RPD	Reactive Power Device
RTU	Remote Terminal Unit
SC	Secondary Control
SCADA	Supervisory Control and Data Acquisition
SE	State Estimator
SGU	Significant Grid User
SM	Synchronous Machine
SmaGPa	Finding the Smart Grid Paradigm and New Architecture to enhance the Controllability associated with Future Power System Operation
SMPS	Switch-Mode Power Supply
SO	System Operator
SSSC	Static Synchronous Series Compensator
STATCOM	Static Synchronous Compensator
StO	Storage-Link Operator
STR	Supplying Transformer
SVC	Static Var Compensator
SyC	Synchronous Condenser
TCSC	Thyristor-Controlled Series Capacitor
TCSR	Thyristor-Controlled Series Reactor
TSO	Transmission System Operator
TSSC	Thyristor-Switched Series Capacitor
TSSR	Thyristor-Switched Series Reactor
TVPP	Technical Virtual Power Plant
UML	Unified Modelling Language
VI	Violation Index
VMD	Voltage Maintaining Devices
VPP	Virtual Power Plant
VSC	Voltage-Source Converter
VvSC	Volt/Var Secondary Control
WoC	Web of Cells
WSC	Watt Secondary Control
WT	Wind Turbine

Graphical Symbols Used in *LINK*-Solution

Graphical Symbol	Name	Description
Dev.	Consuming device	Converts electricity into service for the end-user
Dev.	Devstorage	Consuming device with energy storage potential
→	One directional arrow of active power	The directional arrow of active power is coloured red
→	One directional arrow of reactive power	The directional arrow of reactive power is coloured blue
↔	Bidirectional arrow of active power	
↔	Bidirectional arrow of reactive power	
⬯	*LINK*	It represents the *LINK*–Paradigm and the architectural elements such as grid, producer and storage links
#	Link grid	It represents a grid area. The power flows are bidirectional: in or out of the grid area
↔ ↔	Loadinj	It presents the dual behaviour of a grid part like a **load** or **injection** → Loadinj
St	Storage	Storage in general
St-C ↘ or St ↘	Storage category A	The stored energy is injected at the charging point of the grid, such as pumped hydroelectric storage and stationary batteries

(continued)

(continued)

Graphical Symbol	Name	Description
St-B → or St →	Storage category B	The stored energy is not injected back at the charging point on the grid, such as the power to gas and batteries of e_cars
St-C ↘ or St ↘	Storage category C	The stored energy reduces the electricity consumption at the charging point in the near future, such as cooling and heating systems (consuming devices with energy storage potential)
HzWSC	Hz/Watt secondary control	It maintains a balance between generation and consumption (demand) within the control area and the synchronous area's system frequency
VvSC	Volt/Var secondary control	It maintains the voltage in the whole control area within the limits and the exchanged reactive power at the boundaries according to the schedules

Chapter 1
Introduction*

> Each time we get into this logjam of too much trouble, too many
> problems, it is because the methods that we are using are just
> like the ones we have used before.
> —Richard Feynman

In the last 20 years, many papers and books developed to investigate and design Smart Grids have been written. Smart Grid concepts are introduced, and various models are developed that lead to highly ramified and complex schemas. Different studies have focused on specific parts of power systems regardless of Smart Grids' integrity, leaving all efforts at the level of prototypes or isolated model regions. Results diverge instead of converging towards a complete Smart Grid solution.

So, what is the problem? "An extremely diverse and complex topic" is the answer. "… Each time we get into this logjam of too much trouble, too many problems, it is because the methods that we are using are just like the ones we have used before …" teaches us Richard Feynman [24].

The problem may have its origin in the definitions of the circulating Smart Grid concepts. Without elaborate concepts or paradigms characterised by unique and independent elements, serious design flaws and unclear operating procedures may result [43]. In this chapter, the scope of smart grids is specified, followed by an analysis of the state of the art of the most popular smart grid concepts such as Virtual Power Plants, Microgrids, etc. A descriptive presentation of power systems and the philosophical principles underlying the book are also given.

---

*Author: Albana Ilo

The original version of this chapter was revised: Figure 1.7 has been corrected. The correction to this chapter is available at https://doi.org/10.1007/978-3-030-81530-1_5

© Springer Nature Switzerland AG 2022, corrected publication 2022
A. Ilo and D.-L. Schultis, *A Holistic Solution for Smart Grids
based on LINK–Paradigm*, Power Systems,
https://doi.org/10.1007/978-3-030-81530-1_1

1.1 Smart Grid Scope

Historically, power systems comprised the grid and the power plants: their extension ended at the points of connection with the consumers. The latter were treated as loads with individual behaviour. Figure 1.1a shows the scope of traditional power systems. It includes the grids of all voltage levels, i.e. High (HV), Medium (MV), and Low Voltage (LV) and the power plants mainly connected to the HV grid. Electricity storages such as pumped hydroelectric plants were classified as power plants. The number of power plants connected to the MV grid was minimal.

After the blackout that plagued the United States and Canada, on August 14th, 2003, the term "Smart Grid" was introduced. It was mostly related to the increase of the transmission capacity and level of automation in the grid [13]. The electricity crisis in California (2000–2001) gave the rise of Distributed Generation (DG) a significant boost [53]. The small Photovoltaic (PV) plants on house roofs transformed consumers during the day into electricity producers. A new category of customers appeared: the prosumers.

Over time, the meaning of Smart Grids term has evolved and now stands for modernising power systems and meeting all the requirements. There are several Smart Grid definitions ([51] European Commission; Smart Grid Mandate 2011; [3, 10]). According to [51], Smart Grids' area covers the entire power system right down to the individual electrical appliances in Customer Plants (CP). Figure 1.1b shows the scope of Smart Grids. It includes central and distributed generations connected across the entire power grid (in HV, MV, and LV levels) and customer plants. Smart Grids extend to the electrical devices in the customer plants.

1.2 State of the Art

The power industry has been challenged more than ever in the past 20 years. Its liberalisation [4] and the establishment of the electricity market [14] gave the power industry the first blow, followed by other difficulties caused by introducing new Renewable Energy Resources (RES) and DGs. At present, many countries are moving towards a renewable energy portfolio [15, 17, 23, 25], which increases the fluctuations in power systems. The electricity supply structure is also changing drastically due to many DG units [1, 6, 22, 53] interfering with the system operation at all voltage levels. This kind of development of the power industry is causing severe problems in the management and use of the existing transmission [33, 41, 44, 61] and distribution [27, 35, 55] grids.

To solve those problems, various Smart Grids concepts such as:

- Virtual Power Plants, VPP,
- Microgrids,
- Cellular Approach, and
- Web of Cells (WoC)

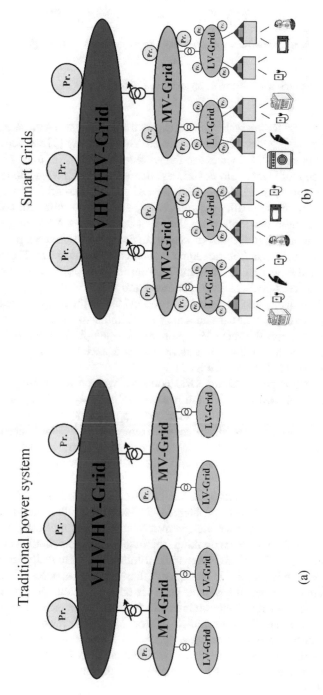

Fig. 1.1 Schematic representation of the scope of: **a** Traditional power system; **b** Smart Grids

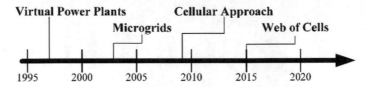

Fig. 1.2 The appearing timeline of the most popular Smart Grid concepts

have been developed for years. Figure 1.2 shows the timeline when these concepts first appeared in the literature. The book entitled Virtual Utility [2] introduced the terminology VPP to make DGs visible in the wholesale market. It quickly became clear that the concept of VPP could not address the technical challenges in terms of frequency and voltage caused by increasing the DG share: Microgrids concept was emerged to overcome these challenges [39]. Many scientists in different countries have started to use these concepts since their first introduction, providing various solutions to their cases. Most of them succeeded in changing these concepts' fundamental definitions to meet the specific requirements set in their projects. The detailed definitions of Virtual Power Plants [52] and Microgrids [32] concepts are still being discussed in various technical-scientific forums worldwide.

The Cellular Approach concept was introduced in 2009 in the research area Smart Grid [7, 12, 37]. A simple comparison shows that similar to Microgrids [62], the Cellular Approach [7] uses the same Matryoshka-doll principle in cells' settings. As emphasised in [38], both concepts are intertwined, and since then, the mixed name of "Smart Micro Grid Cell" was introduced.

In 2015, the landscape of Smart Grid concepts was enriched with the new WoC concept. The cell definition, which has evolved very dynamically in a short time [56, 18, 48, 56], is generic and does not appear unique in the literature [9, 28]. In the cell-based decentralised control framework WoC, cells are defined as "non-overlapping topological subsets of a power system associated with a scale-independent operational responsibility" [28]. Prosumers are not considered explicitly.

To solve the challenge of DG integration, the introduction and use of Smart Grid concepts have caused many other problems [45, 54, 60, 62]. The proposed solutions mainly were of a partial character, in many cases very complicated and with extremely ramified control schemes, making their practical implementation on a large scale almost impossible. Therefore, none of the solutions has practically crossed the boundaries of a research project. Unacceptable these problems, accumulated over time, sooner or later endanger the reliable and safe power systems' operation.

Additionally, the need for a new architecture is noted in the scientific community very early. The Smart Grid Architecture Model (SGAM) introduced in 2012 comprises a framework for the unique description of system architectures for Smart Grids (CEN-CENELEC-ETSI Smart Grid Coordination Group 2012). Its use is very complex and not broad enough for the new emerging requirement of decarbonisation of the economy [21].

Amidst such a research landscape with so many different concepts and restricted architectural model, it appears intriguing to pursue the question of whether the already existing concepts or their combination are likely to support a complete solution for Smart Grids. It is to be expected that any future extension of them or the development of the new concepts should enable the development of a complete Smart Grid solution.

1.2.1   Evaluation of the Most Popular Smart Grids' Concepts

Based on [51] Smart Grid definition, the complete Smart Grid solution is an answer seeking to solve the Smart Grid problems as a whole.

> A complete Smart Grid solution should guarantee a stable, reliable, and cost-effective operation of a more environmental-friendly smart power system. It should also have the ability to ride through the transition phase and further without causing any problems.

1.2.1.1   Evaluation Methodology

The evaluation system consists of two parts: the assessment criteria and evaluation cloud chart.

Assessment criteria.

A complete Smart Grid solution is measured by assessment criteria that consist of a set of benchmarks or yardsticks against which the accomplishment, conformance, performance, and suitability. They are established by considering the following properties: each of them should be unambiguous, comprehensive, direct, operational, and understandable [34].

Ten unique criteria are defined to evaluate the Smart Grids' concepts (see A.1.1):

1.   All voltage levels of power grids.
2.   All electricity producers, regardless of size and technology.
3.   All electricity storages, regardless of size and technology.
4.   Consumers and prosumers.
5.   All power system operation processes.
6.   Data privacy protection.
7.   Avoidance of big data transfer.
8.   Market aspect.
9.   Standardised structure.
10   Transition process.

Fig. 1.3 The evaluation cloud-chart of an ideal Smart Grid solution. All criteria are fulfilled as a whole

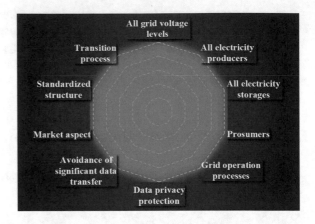

Evaluation cloud chart

Evaluation cloud charts are used to present the assessment results to facilitate their interpretation and compare different cases. Figure 1.3 shows the evaluation cloud-chart of an ideal, complete Smart Grid solution when all criteria are fulfilled as a whole. The study performs only a qualitative analysis and draws results, which should be treated as a trend rather than a sharp rating.

1.2.1.2 Assessment of Smart Grids' Concepts

The detailed assessment for each Smart Grid concept is given in appendix A.1.2 Assessment of popular Smart Grid concepts. Figure 1.4 shows the evaluation cloud charts of each Smart Grid concept.

It shows that none of the solutions fulfill all the evaluation criteria as a whole. They all show almost the same shape, with a significant difference that the grid consideration is successively increased from the VPP solution, Microgrid, Cellular Approach and then up to the Web of Cells. The evaluation cloud-charts are not complementary to each other, which means combining solutions based on various existing Smart Grid concepts, VPP, Microgrids, Cellular Approach, and Web of Cells, cannot provide a complete Smart Grid solution.

1.2.2 Why Cannot Work the Actual Concepts

By examining the most popular Smart Grid concepts, it was found that they cannot characterise the manifold and the complexity of Smart Grids in their entirety. Their definitions are still under development and have not reached the final form. The derived architectural models are pretty complex and challenging to achieve, unlikely to be saved by downstream technical measures.

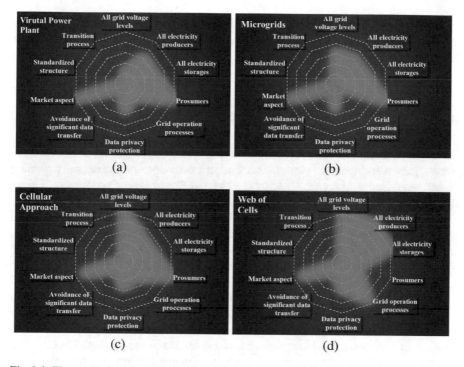

Fig. 1.4 The evaluation cloud-chart of each Smart Grid concept: **a** Virtual Power Plants; **b** Microgrids; **c** Cellular approach; **d** Web of Cells

The cause of the problem lies in origin, in the definitions of the popular Smart Grid concepts. The new architecting theory emphasizes that concepts or paradigms with unique and independent elements are needed to avoid serious design flaws and unclear operating procedures [43]. For example, all Smart Grid concepts that include loads in their definition do not meet the basic principles of architecture theory to have unique elements because the term "load" can have several meanings in power system engineering. According to [30, 31, 47], the load may represent:

- A device connected to a power system that consumes power;
- The total power (active and reactive) consumed by all devices connected to a power system;
- A system portion that is not explicitly represented in a system model but is treated as if it were a single power-consuming device connected to a bus in the system model; and
- The power output of a generator or generating plant.

In addition to the term "load," the Microgrid definition also uses the ambiguous term "host power system;" Experts may interpret it differently.

The popular Smart Grid concepts do not fulfil the principle of the modern architecting theory. Their definitions are not complete and imply elements that are not
unique and independent.

1.2.3 Interdependent and Ambiguous Dimensions of SGAM

The SGAM framework and its methodology are intended to present Smart Grid use
cases from an architectural viewpoint, allowing it both- specific and neutral regarding
solution and technology (CEN-CENELEC-ETSI Smart Grid Coordination Group
[16]). It should enable the validation of Smart Grid use cases.

The SGAM framework consists of five layers representing business objectives and
processes, functions, information exchange and models, communication protocols,
and components. Each layer covers the Smart Grid Plane, spanned by electrical
domains and information management zones (see A.1.3).

The modern architectural theory postulates independent and unique fundamental
elements for a successful design [43]. Figure 1.5 shows an artistic representation
of essential architectural elements through LEGO building blocks. The independent
and unique elements depicted in Fig. 1.5a allow for a solid and unique architecture
design. In contrast, the interdependent and ambiguous elements shown in Fig. 1.5b
provokes crossing overs and often leads to significant problems up to ambivalent,
highly complex architecture designs or no solution at all.

The Smart Grid Plane has a matrix structure (see A.1.3) with two interdependent
and ambiguous dimensions (Domains and Zones). A similar interdependent and
ambiguous nature is also present by the sub-dimensions. For example, on the one
hand, DERs representing distributed generation and storage form a sub-dimension
or domain. On the other hand, Customer Premises, which includes distributed
generation (roof-PV facilities) and storage (battery of e-cars), are another sub-
dimension or domain in the Domains dimension. Both sub-dimensions, DERs and
Customer Premises, are closely intertwined, making these sub-dimensions dependent on each other. Similar interdependences and ambiguities may be found for each
sub-dimension of the Domains and Zones dimensions. SGAM does not support the

Fig. 1.5 Artistic
representation of essential
architectural elements
through LEGO building
blocks: **a** Independent and
unique elements; **b**
Interdependent and
ambiguous elements
(drawing S. Lengauer)

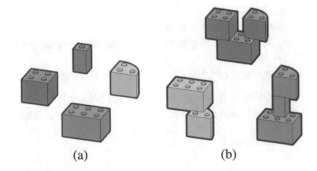

(a)   (b)

various storage options such as chemical storage, P2G, etc. (SGAM [16]) and should be extended beyond the original application to meet the time requirements [21].

1.3 Change of Perspective

Let's look at the schematically represented object in Fig. 1.6. It appears as a rectangle in the profile and horizontal plane and a circle in the vertical one. That is to say, in the two-dimensional perspective, different parties perceive the same object differently. The discussion on the "essence of the object" is therefore inevitable.

To get an objective picture of the whole situation, people first have to exchange their knowledge without prejudice, understand the other's arguments, and accept them—if there are no mistakes. In many cases, we may have to turn to an entirely new, unfamiliar point of view. In the picture, this is a change from the two- to the three-dimensional one. The discussed object is rather a square nor a circle. It is a cylinder.

Change perspective

2D ⟹ **3D**

This fact is an essential indication that a new perspective opens new horizons for a rational solution.

Let us extend the discussion to the area of Smart Grids, where for more than 15 years, controversial discussions have been flooding the scientific literature around

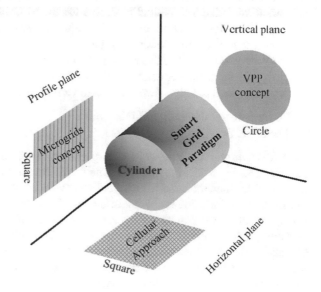

Fig. 1.6 Depending on the perspective, the Smart Grid concept is viewed and designed in entirely different ways

the world. Figure 1.6 also depicts the most popular Smart Grids concepts: Virtual Power Plants, Microgrids, and Cellular Approach. VPP concept is visualised by a circle put on the vertical plane because it was developed considering a partial aspect of Smart Grids: enabling the holder of distributed generators in the market. The microgrid concept is visualised by a square put on the profile plane because it was developed taking into account the partial technical aspect of distribution grids with Distributed Energy Resources (DER) leaving the transmission grid and the large power plants undefined. The Cellular Approach is also visualised by a square put on the horizontal plane. It attempts to further develop the Microgrids concept by considering the distribution and transmission grid through "connection and transmission corridors," respectively. As discussed in the case of the cylinder, the dispute over "the essence of the paradigm" is inevitable. If you want to assess the situation fully, you may need to change the perspective. This change is not always as easy as it seems. "Go to the other side and look from there" is only the first step. Again, we may need to turn to an entirely new, unknown perspective to define the Smart Grid paradigm. It requires a perspective change from the partial to the holistic view.

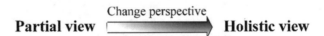

Partial view   Change perspective   **Holistic view**

During this phase, a superordinate connection between some well-known operation processes as "Load/frequency control in high voltage grids" and "Volt/var control in medium voltage grids" was established: The holistic technical model was predicted as shown in Fig. 1.7. According to the prediction, the holistic technical model should include the entire power system, i.e., high-, medium- and low voltage levels, and the customer plants. In the SmaGPa project, scientists initially used the bottom-up or induction method to derive the Smart Grid paradigm *LINK* and the corresponding holistic architecture. The latter has shown the fractality feature of similar structural details. Misinterpretations or -decisions are pre-programmed in the bottom-up method. For this reason, the bottom-up approach was complemented by the top-down or the deduction method.

Scientists used the top-down method to verify the authenticity of:

- The architectural paradigm *LINK*, and
- The *LINK* Solution,

and to exclude any suspicion or misinterpretation. After a detailed analysis, the fractal pattern of Smart Grids was revealed. The *LINK*-Paradigm is derived from the fractal pattern. After that, the holistic technical and market model is conceived, followed by the *LINK*-based holistic architecture design. The latter enables the description and realisation of all processes required for the operation of Smart Grids, such as demand response, load/generation balance, and so on. The holistic architecture merges the electricity producers and storages (regardless of technology or size), the grid (regardless of voltage level), the customer plants, and the market into one structure without compromising data privacy and cybersecurity.

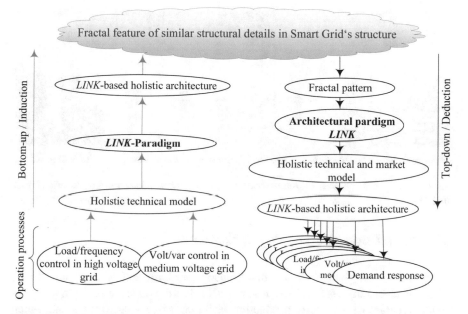

Fig. 1.7 Overview of the bottom-up and top-down methods used to find the *LINK*-Paradigm and design the corresponding holistic architecture

The predictions are confirmed, so we can assume that the holistic architecture is valid under the given circumstances and therefore useable for the practical implementation as shown in this book. The definition of the architectural paradigm of Smart Grids should be clear and contain unique elements so that experts or users cannot make different interpretations. Moreover, Smart Grids, including the entire power system, i.e., from the central and distributed electricity producers and storage through the transmission and distribution grid, and the electrical appliances in the customer plants, i.e., electrical devices, electricity producers (e.g., rooftop PV systems, and so on), and electricity storages, (e.g., the battery of e-cars, and so on) act as a single electromagnetic machine.

The consideration of the Smart Grids' integrity for the architectural design is necessary to achieve reliable and sustainable solutions. Otherwise, in a large-scale implementation, one-sided treatments such as today's battery assignment to a PV system can lead to ineffective solutions with severe environmental consequences such as the formation of mountains with non-recyclable batteries.

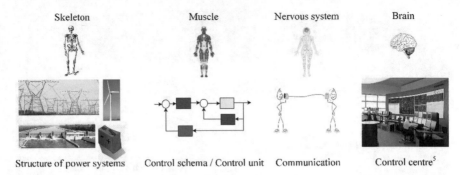

Fig. 1.8 The analogy between the human body and power systems (Control Centre Austrian Power Grid AG)

1.4 Vivid Perception of Power Systems

The world's most giant machines that humans ever built delivered around 25,721 TWh [29] electricity via the power grids in 2017. It is the power system, an interconnected structure of electricity production facilities, wires, transformers, and much more, that keeps the lights on for the homes, buildings, streets, offices, factories, etc. They are very complicated and impossible to model in laboratories; experiments with the physical systems are in almost every country prohibited by law. These features make power systems unique in the landscape of today's technologies; their holistic view is quite complicated.

Figure 1.8 shows a vivid perception of power systems that were crucial for designing the *LINK*-based holistic architecture. It is based on the analogy between the human body and power systems as follows:

- The interconnected structure of electricity production facilities, wires, transformers, and so on is viewed as the skeleton of the human body;
- The control units are considered as muscles, which for example, can close and open the water or gas injectors to increase or decrease electricity production and so on;
- Communication, which includes measurements, is viewed as the nerve system, which contains receptors, that carries information or instructions from the brain to the muscles; while,
- The control centre is considered the brain that controls and operates power systems reliably and safely.

1.5 Philosophical Principles Underlying This Book

Achieving a solution to many of the practical challenges of today's society requires a philosophical statement. In terms of genetic engineering, artificial intelligence, environment or climate: Everywhere, norms and considerations, pros and cons are essential, thus creating the ideal conditions for philosophical discussions and reflections. It is now clear that the new electricity era is a technical, social, and political issue.

The issue of Smart Grids raises many questions, such as the de-carbonisation of the power industry, the role of consumers, the data privacy, the decentralised or centralised solutions, the modernisation or digitalisation of the power grid, etc., which require a clear statement to promote the appropriate technical solution. The questions have been widely discussed in the scientific fore worldwide, but opinions and recommendations remain very different.

The following principles have served as a common thread for the design of the technical solution presented in this book:

- Overall, you would set a sign in total, although your contribution to saving CO_2 could be minimal: It is much better than doing nothing. This principle is reflected in the large-scale integration of distributed renewable energy resources that drive the power industry's de-carbonisation in Smart Grids;
- Reducing the ecological footprint is a necessity to enable the continuity of human life on earth. In terms of Smart Grids, this is reflected in the efficient use of the power industry's existing infrastructure:
- The privacy of the individual, customer, or company is inviolable. In terms of Smart Grids, this is reflected in the reduction of the technical solutions' information;
- The technical solution should guarantee freedom of choice for every customer and should not be discriminatory.

1.6 Outline of the Book

The second chapter gives an overview of the methodology used in this book. The fractal structure's identified signature of Smart Grids, the so-called Smart Grids' fractal pattern, constitutes the *LINK*-Paradigm's foundation. The derived holistic model and architecture are broadly presented. Additionally, it is given a compact description of the control chain strategy and some processes such as the market participation harmonised with the technical holistic architecture, demand response, etc. The third chapter is dedicated to the Energy Systems Integration by treating the Sector Coupling and Energy Communities in the *LINK*-Solution context; meanwhile, the coupling with the non-energy sectors will be treated in another edition of this book. Chapter four deals with the Volt/var process of the control chain over the high-, medium- and low voltage levels and customer plants. The traditionally, recently used and *LINK*-chained control strategies are analysed and developed.

Appendix

A.1.1 Assessment Criteria

The evaluation criteria of a complete Smart Grid solution have to consider the following:

(1) All voltage levels of power grids

Based on [51], both the transmission (very high and high voltage level) and distribution (medium and low voltage level) grid are included in the Smart Grid definition. And rightly so, although different parts of the grid are operated from various utilities, the power grid is one unique giant physical system with very complicated interdependencies [40, 51] between different parts of it. The isolated consideration of particular grid parts may challenge the reliable operation of the neighbouring ones. As reported in (Per [41], the DG's local control connected in a medium voltage grid has created a remarkable uncontrolled reactive power flow in the high voltage grid. Therefore, it is indispensable to consider all voltage levels of power grids for a suitable and complete Smart Grid solution.

(2) All electricity producers, regardless of size and technology.

Based on [51], all electricity producers from the centralised up to the DGs are included in the Smart Grid definition. Apart from the big power plants, the distributed generators are gaining in popularity. They are small-sized, based on different technologies, and can be connected across all distribution grids. Like the large power plants, they show considerable flexibility, but they can adversely affect the network operation when they operate in an uncoordinated way. Therefore, considering all electricity producers regardless of their size and technology is a cornerstone for a successful and complete Smart Grid solution.

(3) All electricity storages, regardless of size and technology.

Historically, large pumped hydroelectric storage have been widely used for electricity storage. Nowadays, driven by advanced technologies, smaller storage units are used. These units are connected across all distribution grids up to the customer plants and contribute to their reliable operation. Thus, considering all storage devices regardless of their size and technology opens up great perspectives for the grid's economic operation and effective use of power supplies' installed capacities.

(4) Consumers and prosumers

Nowadays, many consumers have changed their behaviour, mainly because they have installed PVs on their roofs. These consumers, now transformed into prosumers, can consume and inject power into the power grid and deliver flexibility for its operation.

(5) All power system operation processes.

It is essential to guarantee a secure, reliable, and economic operation of the smart power system in the future. The complete Smart Grid solution must support all the processes needed to operate a power system, i.e., power system monitoring, limit assessment, static and dynamic security, congestion management, outage management, load shedding management, generation load balance, reserve management, restoration, demand response, etc. [64].

(6) Data privacy protection.

Data privacy gets a great relevance only when the prosumers actively participate in the grid operation. The prosumer participation in the grid operation leads to an increase in the use of the information dealing with their private consumption and daily personal activities.

(7) Avoidance of significant data transfer.

Electricity is generally taken for granted and should be available on demand. In many cases, the deployment of the existing Smart Grid solutions leads to a significant data transfer. The latter requires the integration of new Information and Communication Technologies (ICT), which are primarily exposed to cyber-attacks and thus reduce supply reliability. The minimisation of data transfer and secure communication media are crucial for a reliable and secure power supply.

(8) Market aspect.

The electricity market is already an integral part of the power industry. But there are participating almost big power plants. The DGs participation in the market is an immense challenge of today. Therefore, market flourishing is essential for the development of Smart Grids [3].

(9) Standardised structure

Standardised structures can be easily deployed into the power industry and contribute to the economic, qualitative, simple, and straightforward development and implementation of a given solution. They are fundamental to the large-scale implementation of the complete Smart Grid solution.

(10) Transition process

Electricity is one of the fundamentals of today's life because we depend on electricity for such rudimentary needs as water, food, health care, communications, etc. The electricity industry is classified as one of our society's most critical infrastructure [50]. This classification means that smart power systems should operate reliably during their transition time as well as afterwards. Therefore, the availability of a smooth transition process from the existing power systems and applications to the Smart Grids and the corresponding applications is an indispensable criterion for the complete Smart Grid solution.

A.1.2 Assessment of Popular Smart Grid Concepts

A.1.2.1 Virtual Power Plants

Figure 1.9 shows the cloud chart for the Smart Grid solution based on the VPP concept. It is composed of several DERs of various technologies with various operating patterns and availability; the characteristics of the VPP may vary significantly in time. This concept was initially designed to enable the small-decentralised electricity producer to participate in the market. Later, it was found out that the complex management processes of the grid could not be considered with this concept [19]. Therefore, the VPP concept was split up into the Commercial VPP (CVPP) and the Technical VPP (TVPP) [57]. The CVPP represents a portfolio of DERs that can be used to participate in energy markets in the same manner as the transmission-connected generating plant. The TVPP provides DER visibility to the system operator(s); it allows DERs to contribute to system management activities and facilitates the use of DER capacity, providing system balancing at the lowest cost. The power grid is not treated explicitly.

Therefore, the first criterion that requires that all voltage levels of power grids are taken into consideration seems not to be met. All studies based on the VPP consider the distributed generation and storage [5, 8, 19, 36, 52, 54]. The large power and storage plants connected to the high voltage grid are not taken into account. Criteria 2 and 3, which consider all producers and storages, regardless of size and technology, are not fulfilled as a whole. Consumers and prosumers are fully taken into account; thus, criterion 4, requiring that consumers and prosumers are taken into consideration, is met in its entirety. Criterion 5 considers all power system operation processes. The generation-balance process is being broadly treated. Other

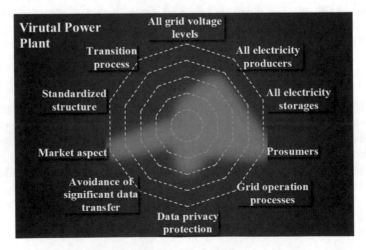

Fig. 1.9 The evaluation cloud-chart of Virtual Power Plants

grid-related operating processes [64] are not treated or are still part of the research process in the context of the TVPP. Therefore, criterion 5 can hardly be fulfilled in this case. Solutions based on the VPP require a significant data exchange between the DERs and the aggregators and Transmission (TSO) and Distribution System Operators (DSO) [57], thus creating an ICT challenge and data privacy issue [41, 44]. Much research has been carried out in this area, thus, criterion 6 on data privacy protection results to be partially fulfilled. Criterion 7, which requires the avoidance of significant data transfer [57], is not fulfilled. Criterion 8, which deals with the market aspect, is met in its entirety. All solutions derived from the VPP are very intricate with extremely ramified schemes [11, 26, 41, 44, 49, 63]. No standardised structure could be identified, and, as a result, criterion 9 is not met. Criterion 10 on the transition process is not dealt with in literature, so it is not fulfilled.

A.1.2.2 Microgrids

Figure 1.10 shows the cloud chart of the Smart Grid solution, which is based on the Microgrids concept. Although a detailed definition of Microgrids is still under discussion in technical forums, a Microgrid can be described as a cluster of loads, distributed generation units, and energy storage systems operated in coordination to reliably supply electricity, connected to the host power system at the distribution level at a single point of connection, the point of common coupling [32]. Due to the definition, the power grid cannot be treated in its entirety. Microgrids apply only to distribution, which means medium and low voltage grids. The high voltage grid may be included in the "host power system" [32], "Macrogrid" [59], which per definition is not part of Microgrid. Therefore, the first criterion that requires that all voltage

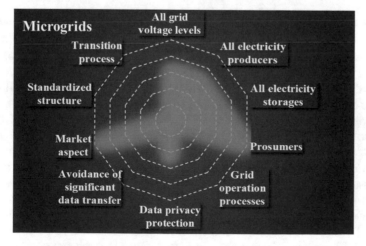

Fig. 1.10 The evaluation cloud-chart of Microgrids

levels of power grids are considered is not met in its entirety. Distributed generation and storage are a fundamental part of Microgrids, but the ones connected to the high voltage grid are not taken into account. Consequently, criteria 2 and 3, which consider all producers and storages regardless of size and technology, are not fulfilled as a whole. Consumers and prosumers are fully taken into account; thus, criterion 4, requiring that consumers and prosumers are considered, is met in its entirety. Criterion 5 deals with the consideration of all power system operation processes. Although a Microgrid should be capable of operating in grid-connected and stand-alone modes [32], no contribution can be found that describes the grid-connected regime. As a result, the power system operation processes are not treated. In these conditions, the 5th criterion can hardly be fulfilled.

Data privacy is a crucial issue in the Microgrids solutions when there is needed cooperation among subsystems [45]. A lot of research work has been carried out in this area [59], thus, criterion 6 on data privacy protection results to be partially fulfilled. Microgrids solutions require significant data transfer, thus creating an ICT challenge [59]. Additionally, the game theory usage for the Smart Grid solution increases the risk of cyber-attacks drastically. Here a lot of research has been carried out, hence criterion 7, which requires the avoidance of significant data transfer results somewhat fulfilled. Criterion 8 deals with the consideration of the market aspect. Many attempts are made to implement a market environment using a Multi-agent System [20] or a market model based on Game Theory [65]. Therefore, criterion 8 can be considered fulfilled. No standardised structure could be identified, and as a result, criterion 9 is not met. Criterion 10 on the transition process is not dealt with in the literature, so it is not fulfilled.

A.1.2.3 Cellular Approach

Figure 1.11 shows the cloud chart for the Smart Grid solution, which is based on the Cellular Approach. It can be seen as a self-controlled small microgrid integrated with a modular Smart Grid ICT infrastructure [23]. The distribution grid is considered through "connection corridors" while the transmission grid through "transmission corridors" [7]. Therefore, the first criterion, which requires considering all voltage levels of the power grid, is fulfilled in its entirety. The Cellular Approach applies to the distributed producers and storages [7]. Still, the big power plants and storages connected to the high voltage grid are not considered. Criteria 2 and 3, which consider all producers and storages regardless of size and technology, are not fulfilled as a whole. Consumers and prosumers are fully considered; thus, criterion 4, requiring that consumers and prosumers are taken into account, is met in its entirety. Criterion 5 deals with the consideration of all power system operation processes [64]. Cellular Approach deals mainly with the process of load-generation balance combined with those of Demand Side Management. Other essential operations [64] are, in fact, not addressed. Therefore, criterion 5 seems to be fulfilled to some extent.

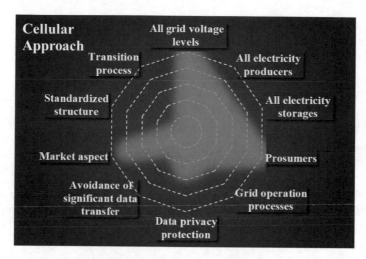

Fig. 1.11 The evaluation cloud-chart of the Cellular Approach

Data privacy, criterion 6, and the avoidance of significant data transfer, criterion 7, are the biggest challenges in realising the Cellular Approach. Indeed, researchers have concluded that further research projects and concepts are needed to develop robust solutions against cyber-attacks [37]. Therefore, criteria 6 and 7 seem to be little fulfilled. The Cellular Approach supports and requires a new market design and business models, which are already in process [7]. Therefore, criterion 8, which deals with the market aspect, is not fulfilled as a whole. No standardised structure could be identified, and, as a result, criterion 9 is not met. Criterion 10 on the transition process is not dealt with in the literature, so it is not fulfilled.

A.1.2.4 Web of Cells

Figure 1.12 shows the cloud chart for the Smart Grid solution, which is based on the WoC. The latter considers all voltage levels of the power grid, thus fulfilling the first criteria in its entirety.

Also, criteria 2 and 3, which consider all producers and storages regardless of size and technology, are fulfilled as a whole. Consumers and prosumers are not explicitly taken into account; thus, criterion 4 requiring that consumers and prosumers are considered is almost not met. Criterion 5 deals with the consideration of all power system operation processes [64]. Various operation processes are dealt with under the WoC architecture, such as load–frequency control [56, 58] reserve monitoring [28], voltage control [46], and optimisation process [42]. The emergency and price-driven demand response process is missing in the literature. Therefore, criterion 5 seems to be partly fulfilled. Data privacy, criterion 6, and the avoidance of big data transfer, criterion 7, are the biggest challenges in the realisation of the WoC [21]. Therefore,

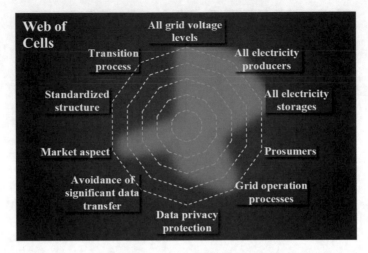

Fig. 1.12 The evaluation cloud-chart of Web of Cells

criteria 6 and 7 seem to be little fulfilled. WoC supports and requires a new market design and business models, which are already in process [9]. Therefore, criterion 8, which deals with the market aspect, is not fulfilled as a whole. No standardised structure could be identified, and, as a result, criterion 9 on the standardised structure is not met. Criterion 10 on the transition process is not dealt with in the literature, so it is not fulfilled.

A.1.3 SGAM Smart Grid Plane

The Smart Grid Plane (CEN-CENELEC-ETSI Smart Grid Coordination Group [16]) is defined from the application to the Smart Grid Conceptual Model of the principle of separating the Electrical Process viewpoint (partitioning into the physical domains of the electrical energy conversion chain) and the Information Management view-point (partitioning into the hierarchical zones (or levels) for the management of the electrical process, Fig. 1.13.

A.1.3.1 SGAM Domain

One dimension of the Smart Grid Plane covers the complete electrical energy conversion chain, partitioned into five domains: Bulk Generation, Transmission, Distribution, DER and Customers Premises (Table 1.1).

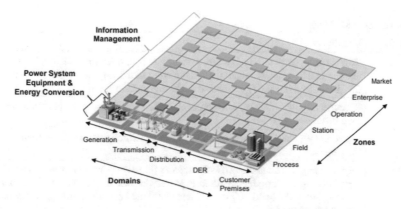

Fig. 1.13 Smart Grid plane—domains and hierarchical zones (CEN-CENELEC-ETSI Smart Grid Coordination Group 2012)

Table 1.1 SGAM Zones

Domain	Description
Bulk Generation	Representing generation of electrical energy in bulk quantities, such as by fossil, nuclear and hydropower plants, off-shore wind farms, large scale solar power plant (i.e. PV, CSP)—typically connected to the transmission system
Transmission	Representing the infrastructure and organisation which transports electricity over long distances
Distribution	Representing the infrastructure and organisation which distributes electricity to customers
DER	Representing distributed electrical resources directly connected to the public distribution grid, applying small-scale power generation technologies (typically in the range of 3 kW to 10.000 kW). These distributed electrical resources may be directly controlled by DSO
Customer Premises	Hosting both—end users of electricity, also producers of electricity. The premises include industrial, commercial and home facilities (e.g. chemical plants, airports, harbours, shopping centres, homes). Also, generation in the form of, e.g. photovoltaic generation, electric vehicles storage, batteries, micro turbines… are hosted

A.1.3.2 SGAM Zone

One dimension of the Smart Grid Plane represents the hierarchical levels of power system management, partitioned into six zones: Process, Field, Station, Operation, Enterprise and Market (Table 1.2).

Table 1.2 SGAM Zones

Domain	Description
Process	Representing generation of electrical energy in bulk quantities, such as by fossil, nuclear and hydropower plants, off-shore wind farms, large scale solar power plant (i.e. PV, CSP)– typically connected to the transmission system
Field	Representing the infrastructure and organisation which transports electricity over long distances
Station	Representing the infrastructure and organisation which distributes electricity to customers
Operation	Representing distributed electrical resources directly connected to the public distribution grid, applying small-scale power generation technologies (typically in the range of 3 kW to 10.000 kW). These distributed electrical resources may be directly controlled by DSO
Enterprise	Includes commercial and organisational processes, services and infrastructures for enterprises (utilities, service providers, energy traders …), e.g. asset management, logistics, workforce management, staff training, customer relation management, billing and procurement…
Market	Reflecting the market operations possible along the energy conversion chain, e.g. energy trading, mass market, retail market

References

1. Ackermann T, Anderson G, Söder L (2001) Distributed generation: a definition. Electric Power Syst Res 57:195–204
2. Awerbuch S, Preston AM (1997) The virtual utility: accounting, technology and competitive aspects of the emerging industry, vol 26, Kluwer Academic Pub
3. BDEW (2013) BDEW Roadmap—realistic steps for the implementation of Smart Grids in Germany. https://www.bdew.de/media/documents/Pub_20130211_Roadmap-Smart-Grids_english.pdf. Accessed 20 Mai 2021
4. Bacon RW (1995) Privatisation and reform in the global electricity supply industry. Annu Rev Energy Env 20(1):119–143. https://doi.org/10.1146/annurev.eg.20.110195.001003
5. Bakari EK, Kling WL (2010) Virtual power plants: an answer to increasing distributed generation. In: 2010 IEEE PES Conference on Innovative Smart Grids Technologies Conference Europe, Gothenburg, 11–13 October 2010. Institute of Electrical and Electronics Engineers, Piscataway, pp 1–7, pp 10–13. https://doi.org/10.1109/ISGTEUROPE.2010.5638984
6. Barker PP, Mello RWD (2000) Determining the impact of distributed generation on power systems. I. Radial distribution systems. In: IEEE Power Engineering Society Summer Meeting, Seattle, 16–20 July, pp 1645–1656. https://doi.org/10.1109/PESS.2000.868775
7. Benz T et al (2015) VDE-Studie "Der Zellulare Ansatz". https://shop.vde.com/de/copy-of-vde-studie-der-zellulare-ansatz. Accessed 11 May 2021
8. Bignucolo F, Caldon R, Prandoni V, Spelta S, Vezzola M (2006) The Voltage Control on MV Distribution Networks with Aggregated DG Units (VPP). In: 41st International Universities Power Engineering Conference, Newcastle-upon-Tyne, 6–8 Sept, vol 1, pp 187–192. https://doi.org/10.1109/UPEC.(2006).367741
9. Bobinaite V, Obushevs A, Oleinikova I, Morch A (2018) Economically Efficient design of market for system services under the web-of-cells architecture. Energies 11:729. https://doi.org/10.3390/en11040729
10. Bollen MHJ (2011) The Smart Grid: adapting the power system to the new challenges. Morgan and Claypool Publishers, San Rafael

11. Bose A (2010) Smart transmission grid applications and their supporting infrastructure. In: IEEE Transactions Smart Grid, vol 1, pp 11–19. https://doi.org/10.1109/TSG.2010.2044899
12. Buchholz B, Nestle D, Kiessling A (2009) "Individual customers" influence on the operation of virtual power plants. In: IEEE Power and Energy Society General Meeting, pp 1–6. https://doi.org/10.1109/PES.2009.5275401
13. Burr MT (2003) Reliability demands will drive automation investments. Fortnightly Magazine, pp 1–4. https://www.fortnightly.com/fortnightly/2003/11/technology-corridor. Accessed 03 May 2016
14. Cardell JB, Hittb CC, Hogan WW (1997) Market power and strategic interaction in electricity networks. Resource Energy Econ 19(1–2):109–137. https://doi.org/10.1016/S0928-765 5(97)00006-7
15. Chu J (2015) China's fast track to a renewable future. The climate group RE100 China analysis. https://www.theclimategroup.org/sites/default/files/archive/files/RE100-China-analysis. pdf. Accessed 19 January 2019
16. CEN-CENELEC-ETSI Smart Grid Coordination Group (2012) Smart Grid reference architecture. https://ec.europa.eu/energy/sites/ener/files/documents/xpert_group1_reference_architect ure.pdf Accessed 11 May 2018
17. Clean Energy Council (2015) Clean Energy Australia Report. https://www.cleanenergycoun cil.org.au/policy-advocacy/reports/clean-energy-australia-report.html. Accessed 20 Jan 2018
18. D'hulst R, Fernandez JM, Rikos E, Kolodziej D, Heussen K, Geibelk D, Temiz A, Caerts C (2015) Voltage and frequency control for future power systems: the ELECTRA IRP proposal. In: International symposium on smart Electric Distribution Systems and Technologies (EDST), Vienna, 8–11 September, pp 245–250. https://doi.org/10.1109/SEDST.2015.7315215
19. Dielmann K, Veiden A (2003) Virtual power plants (VPP)—a new perspective for energy generation?. In: 9th international scientific and practical conference of students, Post-Graduates Modern Techniques and Technologies, Tomsk, 7–13 April, pp 18–20
20. Dimeas AL, Hatziargyriou ND (2005) Operation of a multiagent system for microgrid control. IEEE Trans Power Syst 20(3):1447–1455. https://doi.org/10.1109/TPWRS.2005.852060
21. ETIP SNET (2019) White Paper Holistic architectures for future power systems. https://www. etip-snet.eu/white-paper-holistic-architectures-future-power-systems. Accessed 12 April 2019
22. El-Khattam W, Salama MMA (2004) Distributed generation technologies, definitions and benefits. Electric Power Syst Res 71:119–128. https://doi.org/10.1016/j.epsr.2004.01.006
23. European Commission (2016) Directive of the European parliament and of the council on the promotion of the use of energy from renewable sources. https://ec.europa.eu/energy/sites/ener/ files/documents/1_en_act_part1_v7_1.pdf. Accessed 3 Mai 2018
24. Feynman RP (1994) The character of physical law. Modern Library, New York
25. Fleischmann D (2016) Renewable Energy Was 16.9 Percent of U.S. Electric Generation in the First Half of 2016. Renewable Energy World. http://www.renewableenergyworld.com/ articles/2016/08/renewable-energy-was-16-9-percent-of-u-s-electric-generation-in-the-first-half-of-2016.html. Accessed 30 Oct 2019
26. Grijalva S, Tariq MU (2011) Prosumer-based Smart Grid architecture enables a flat, sustainable electricity industry. Innovative Smart Grid Technologies Conference ISGT, Anaheim, 17–19 January, pp 1–6. http://ieeexplore.ieee.org/stamp/stamp.jsp?arnumber=5759167. Accessed 10 Jan 2017
27. Hidayatullah NA, Paracha ZJ, Kalam A (2009) Impacts of distributed generation on Smart Grid. Int Con Electrical Energy Industrial Electronic Syst Malaysia 7–8:218–221
28. Hu J, Lan T, Heussen K, Marinelli M, Prostejovsky A, Lei X (2018) Robust allocation of reserve policies for a multiple-cell based power system. Energies 11:381. https://doi.org/10. 3390/en11020381
29. IEA (2019) Electricity Information 2019 Statistics report—September 2019. https://www.iea. org/reports/electricity-information-2019. Accessed 16 April 2020
30. IEEE Task Force on Load Representation for Dynamic Performance (1993) Load representation for dynamic performance analyses (of power systems). IEEE Trans Power Syst 8: 472–482. https://doi.org/10.1109/59.260837

31. IEEE Task Force on Load Representation for Dynamic Performance (1995) Standard load models for power flow and dynamic performance simulation. IEEE Trans Power Syst 10: 1302–1313. https://doi.org/10.1109/59.466523
32. IEEE Task Force on Microgrid Control (2014) Trends in Microgrid Control. IEEE Trans Smart Grid 5: 1905–1919
33. Ilo A, Gawlik W, Schaffer W, Eichler R (2015) Uncontrolled reactive power flow due to local control of distributed generators. In: 23th International Conference on Electricity Distribution, CIRED, Lyon, 15–18 June, pp 1–5
34. Keeney RL, Gregory RS (2005) Selecting attributes to measure the achievement of objectives. Oper Res 53(1):1–11. https://doi.org/10.1287/opre.1040.0158
35. Kerber G, Witzmann R, Sappl H (2009) Voltage limitation by autonomous reactive power control of grid connected photovoltaic inverters. In: Conference on Compatibility and Power Electronics, Badajoz, 20–22 May, pp 129–133
36. Kieny C, Berseneff B, Hadjsaid N, Besanger Y, Maire J (2009) On the concept and the interest of Virtual Power Plant: some results from the European project FENIX. Power & Energy Society General Meeting, Calgary, 26–30 July, pp 1–6
37. Kießling A (2013) Beiträge von moma zur Transformation des Energiesystems für Nach-haltigkeit, Beteiligung, Regionalität und Verbundheit. Modellstadt Mannheim (moma). https://www.ifeu.de/energie/pdf/moma_Abschlussbericht_ak_V10_1_public.pdf. Accessed 17 July 2017
38. Kleineidam G, Krasser M, Reischböck M (2016) The cellular approach: smart energy region Wunsiedel. Testbed for Smart Grid, smart metering and smart home solutions. Electr Eng 4:335–340. https://doi.org/10.1007/s00202-016-0417-y
39. Lasseter B (2001) Microgrids [distributed power generation]. IEEE Power Engineering Society Winter Meeting, Columbus, 28 January–1 February, vol 1, pp 146–149
40. Lasseter RH et al (2002) The CERTS Microgrid Concept. White paper for Transmission Re-liability Program, Office of Power Technologies, U.S. Department of Energy, Washington DC, pp 1–27
41. Lund P (2007) The Danish cell project—Part 1: background and general approach. 2007 IEEE Power Engineering Society General Meeting, Tampa, 24–28 June, pp 24–28
42. MacDougall P, Ran B, Huitema GB, Deconinck G (2017) Multi-goal optimisation of competing aggregators using a web-of-cells approach. In: 2017 IEEE PES Innovative Smart Grid Tech-nologies Conference Europe (ISGT-Europe), Torino, 26–29 September, pp 1–6. https://doi.org/10.1109/ISGTEurope.2017.8260335
43. Maier MW, Rechtin E (2009) The art of systems architecting. CRC Press, Taylor & Francis Group, Boca Raton, ISBN 9781420079135
44. Martensen N, Kley H, Cherian S, Pacific O, Lund P (2009) The cell controller pilot project: testing a smart distribution grid in Denmark. Grid Interop, The Road to an Interoperable Grid, pp 216–222
45. McDaniel P, McLaughlin S (2009) Security and privacy challenges in the Smart Grid. IEEE Secur Priv 7(3):75–77. https://doi.org/10.1109/MSP.2009.76
46. Merino J et al (2017) Electra IRP voltage control strategy for enhancing power system stability in future grid architectures. CIRED—Open Access Proc J 1:1068–1072. https://doi.org/10.1049/oap-cired.2017.0749
47. Milanovic JV, Matevosyan J, Gaikwad A, Borghetti A (2014) Modelling and aggregation of loads in flexible power networks. CIGRE, W.G. C4.605, pp 1–191. https://e-cigre.org/publication/566-modelling-and-aggregation-of-loads-in-flexible-power-networks. Accessed 3 Nov 2018
48. Morch AZ, Jakobsen SH, Visscher K, Marinelli M (2015) Future control architecture and emerging observability needs. In: 5th international conference on power engineering, energy and electrical drives, Riga, 11–13 May. IEEE, Piscataway, pp 234–238. https://doi.org/10.1109/PowerEng.2015.7266325
49. Moslehi K, Kumar R (2010) A reliability perspective of the Smart Grid. IEEE Trans Smart Grid 1: 57–64. https://doi.org/10.1109/TSG.2010.2046346

50. Moteff J, Parfomak P (2004) Critical infrastructure and key assets: definition and identification. White Paper, Resources, Science & Industry Division, Library of Congress, Washington DC, Congressional Research Service / Library of Congress. https://fas.org/sgp/crs/RL32631.pdf. Accessed 12 Jun 2018

51. Myrda P (2009) Smart Grid enabled asset management. EPRI, Paolo Alto, Report 1017828

52. Othman MM, Hegazy YG, Abdelaziz AY (2015) A review of virtual power plant definitions, components, framework and optimisation. Int Electrical Eng J 6(9):2010–2024

53. Owens B (2014) The rise of distributed power. General Electric Company, New York, vol 47. https://www.eenews.net/assets/2014/02/25/document_gw_02.pdf Accessed 20 May 2021

54. Plancke G, Vos KD, Belmans R, Delnooz A (2015) Virtual power plants: definition applications and barriers to the implementation in the distribution system. In: 12th International Conference on the European Energy Market, Lisbon, 19–22 May, pp 1–5

55. Prata RA (2006) Impact of distributed generation connection with distribution grids—two case-studies. In: IEEE in power engineering society general meeting, Montreal, 18–22 June, vol 19, pp 1–8, pp 18–22

56. Prostejovsky A, Marinelli MM, Rezkalla M, Syed MH, Guillo-Sansano E (2018) Tuningless load frequency control through active engagement of distributed resources. IEEE Trans Power Syst 33(3):2929–2939. https://doi.org/10.1109/TPWRS.2017.2752962

57. Pudjianto D, Ramsay C, Strbac G (2007) Virtual power plant and system integration of distributed energy resources. IET Renew Power Gener 1:10–16. https://doi.org/10.1049/iet-rpg:20060023

58. Rikos E, Cabiati M, Tornelli C (2017) Adaptive frequency containment and balance restoration controls in a distribution network. In: 2017 IEEE PES Innovative Smart Grid Technologies Conference Europe (ISGT-Europe), Torino, 26–29 Semptember, pp 1–6. https://doi.org/10.1109/ISGTEurope.2017.8260269

59. Saad W et al (2012) Game-theoretic methods for the Smart Grid—an overview of microgrids systems, demand-side management, and smart grid communications. IEEE Signal Process Mag 29(5):86–105. https://doi.org/10.1109/MSP.2012.2186410

60. Schiller C, Fassmann S (2010) The smart micro grid: IT challenges for energy distribution grid operators. White Paper, IBM, pp 36–42. https://www.smartgrid.gov/files/The_Smart_Mcro_Grid_IT_Challenges_for_Energy_Distribution_G_201002.pdf. Accessed 11 Jun 2016

61. Schäfer P, Vennegeerts H, Krahl S, Moser SA (2015) Derivation of recommendations for the future reactive power exchange at the interface between distribution and transmission grid. In: 23th international conference on electricity distribution, CIRED, Lyon, 15–18 June 2015, pp 1–5

62. Soshinskaya M, Crijns-Graus WH, Guerrero JM, Vasquez JC (2014) Microgrids: experiences, barriers and success factors. Renew Sustain Energy Rev 40:659–672. https://doi.org/10.1016/j.rser.2014.07.198

63. Taft J, Martini PD (2012) Ultra large-scale power system control architecture—a strategic framework for integrating advanced grid functionality. https://www.cisco.com/c/dam/en/us/products/collateral/cloud-systems-management/connected-grid-network-management-system/control_architecture.pdf. Accessed 20 May 2021

64. Vaahedi E (2014) Practical power system operation. Wiley, New Jersey

65. Wang Z, Yang K, Wang X (2013) Privacy-preserving energy scheduling in Microgrid systems. IEEE Transactions on Smart Grid, vol 4/4, pp 1810–1820

Chapter 2
Holistic Architecture*

Study the science of art. Study the art of science. Develop your
senses – especially learn how to see. Realise that everything
connects to everything else.
—Leonardo da Vinci

Until recently, the power system's architecture was self-evident to contemporary
engineers' generation; it has had the same architectural shape for decades. Engineers
dealt almost entirely with measurable using analytic tools derived from mathematics,
focusing on technical optimisation and quantifiable costs. Smart Grids' perception
and design, which are extraordinarily complex, exceed the engineers' everyday work
boundaries. It requires a cross-research in the areas of architecture, physics, and
electrical engineering.

Careful observation of nature made it possible to discover fractals, similar patterns,
and shapes that nature repeats in ever-smaller sizes. Fractal analysis is one of the most
advanced and modern methods of today used to expand knowledge about the structures and functions of complex systems or objects in nature, medicine, engineering,
etc. This innovative method is applied to understand and develop the holistic approach
of Smart Grids. A holistic architecture transforms experts' perspectives on a power
system and their perceptions of what might be possible. This new perspective can
help them perceive the power system in more expansive ways and create increasingly
more flexibilities and opportunities to decarbonise the economy.

This chapter presents the first holistic architecture of smart grids comprising the
high-, medium- and low voltage grids, customer plants and the market. It is designed
based on the fractal feature of the grid. The derived *LINK*-paradigm and the corresponding technical and market holistic models are described in detail. The different
architecture levels, operation modes, and new control chain net strategies are some of
the topics treated in this chapter. Additionally, the keystones are given for the chains of
smart grids operation such as monitoring, load/generation balance, Volt/var management, demand response, congestion management, and so on. It also lays out the
fundamentals of smart grid operational chains such as monitoring, load/generation

---

*Author: Albana Ilo

© Springer Nature Switzerland AG 2022
A. Ilo and D.-L. Schultis, *A Holistic Solution for Smart Grids
based on LINK–Paradigm*, Power Systems,
https://doi.org/10.1007/978-3-030-81530-1_2

balancing, volt/var management, demand response, congestion management, etc. Finally, data privacy and cybersecurity, the *LINK* economics and implementation steps are discussed. It does not attempt to gear the holistic approach to the Smart Grid in a particular country. Instead, the basics concepts are highlighted, all of which can be easily implemented to suit each country's Smart Grids' specific circumstances.

2.1 Introduction

The traditional structure of power systems became questionable at the beginning of the liberalisation; the market rules overcome the technical. Several blackouts and electricity crises were the consequence [73]. It quickly became apparent that an ultra-large-scale comprehensive control framework is needed to avoid the growing and unmanageable patchwork of grid implementations that are not sustainable on a large scale [66].

The purpose of architecture is to structure and organise the system in such a way that it is stable, usable, adapt to changes, and economical. When the architecture is sound, it helps to design better the system that is being described. The architecture of a system is a global model of it. It usually consists of a structure, properties of various elements involved, relationships between components, and their behaviour and dynamics.

When we talk about Smart Grids, we must be clear that the giant electromagnetic machine must be treated as a whole. So far, the power system's operation has been treated separately from that of the customer plants and their electrical equipment; their interaction was wasteful. But, the picture changes with the increasing share of DGs. The exchange becomes very strong and cannot be ignored anymore. Therefore, a holistic approach related to or concerned with integrated wholes or complete systems rather than analysing or treating separate parts is indisposable.

The design structure of Smart Grids consists of three complementary and consistent architecture types:

1. The holistic architecture of electrical appliances, including market to enable the processes required for the operation of Smart Grids;

2. The architecture of applications required to run the processes (soft- and hardware architecture); and.

3. Information and communication architecture required for the smooth data exchange between applications.

Whatever the system [50], one of the fundamental principles of system designing is reasoning according to an architectural paradigm. The system can then be linked to another through an interface and viewed on various complementary and consistent abstraction levels.

> The essential requirement of an architectural paradigm is its composition with unique and independent elements.

The analysis made in Sect. 1.2.3 has resulted that none of the popular concepts like VVP, Microgrids, etc. meets the above-mentioned requirement. Hence, the greatest puzzle of our time is identifying the Smart Grid paradigm that meets these requirements.

Last years, the fractal was increasingly being used to describe complex structures in nature [48]. It has recently been considered a valuable tool for analysing, understanding, and designing Smart Grids. According to [42], the scale-invariant properties of power grids are due to their design. Self-similarity is already identified in load flow characteristics [68], power demand behaviour [9], fault diagnostics [47], and protective relays [45, 72]. Similarities have also been discovered in blackouts of interconnected power grids and their subgrids [71]. Fractal principles are studied in grid structures, and fractal dimensions have been calculated for different grids [23, 70] to develop a cluster power system philosophy with cluster network structure [54, 60]. Other authors have organised Smart Grids based on a fractal model by associating it with the holonic concept and holarchy [18]. Others consider fractals as an instrument that facilitates Smart Grids' realisation in a decentralised fashion [51]. However, identifying a fractal pattern, which describes Smart Grids in their entirety, has not yet been a research focus.

This chapter is devoted to designing electrical appliances' holistic architecture, including the market, to enable the processes required to operate Smart Grids. The Smart Grid paradigm, which will be the essential instrument for holistic models and architecture design, is defined based on Smart Grids' fractal pattern.

2.2 Methodology

The definition of the architectural paradigm for Smart Grids should be clear and understandable for everyone. Its elements should be chosen uniquely and independently, without any ambiguity [46]. This way, there will be no possibility for different interpretations, making it easier for experts around the world to work together. Therefore, the fractal pattern of Smart Grids is chosen to form the paradigm's core because, by definition, it contains unique and independent elements. Figure 2.1 shows the methodology used to design the holistic architecture of electrical appliances, including the market. Firstly, an in-depth analysis of Smart Grids' fractal nature is performed to determine their fractal pattern. This analysis is followed by examining Smart Grids' requirements and their combination with the fractal pattern to derive the paradigm. The holistic technical model of Smart Grids is based on the Smart Grid paradigm, while the holistic market model mirrors it. Both the holistic technical

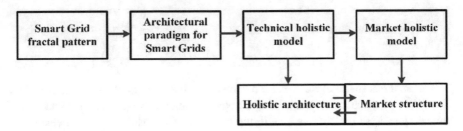

Fig. 2.1 The methodology used to design the holistic architecture

and market models form the heart of the system architecting. Finally, the holistic architecture and associated market structure are designed.

2.3   Fractal in Smart Grids

Fractal is a natural phenomenon or mathematical construct characterised by a cascade of similar structural details that appear at every scale [48]. The cascade is never-ending in theoretical mathematical cases, but it is limited to a finite number of levels in practical cases. The question is whether Smart Grids have a fractal structure. If so, what kind of fractal characterises them? What is their fractal pattern?

2.3.1   Have Smart Grids Fractal Structures?

Fractals can be observed in almost an infinite variety in nature. From the simplicity of a single cell to the immensity of galaxies, fractals are available all over. Flowers, plants, trees, veins in human and animal bodies, hurricanes, etc., have natural fractal structures with different branching or spiral shapes. Like a tree, in which every cell has to be close to a vein to take in water and nutrients, every electrical device in Smart Grids stands close to a conductor to receive electricity. Figure 2.2a shows the fractal branching of an upside-down tree. The tree survives through a fractal branching structure with veins branch and branch ever smaller, down to the width of a capillary. In Smart Grids, the wire is a fundamental component available from the high voltage to the electrical devices at the Customer's Plant (CP) level. Their length and diameter get smaller and smaller with the voltage level as in a branching fractal structure. Figure 2.2b shows the structure of Smart Grids that resembles the fractal branching of an upside-down tree. The corresponding fractal pattern is found out below through a detailed fractal analysis of Smart Grids' structure.

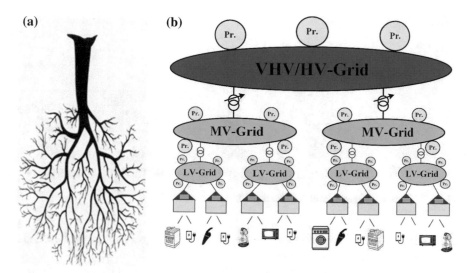

Fig. 2.2 Branching fractal structure in: **a** Upside-down tree; **b** Smart Grids

2.3.2 Fractal Pattern

The search for the fractal pattern of Smart Grids is limited to an ending cascade that starts with the highest voltage level and ending with the customer device level, as shown in Table 2.1. The fractal of Smart Grids is conceived by five fractal levels shown in Table 2.1. To meet the ever-increasing demand for electricity effectively, power engineers have designed and developed power systems based on different voltage levels. The electrical appliances connected in each of the three voltage levels, HV, MV, and LV, are contained in fractal levels 1, 2, and 3. The electrical appliances connected in the CP grid correspond to level 4. Level 5 considers all internal elements of the electrical Devices (Dev) connected by wires/conducting

Table 2.1 The fractal structure of Smart Grids based on different fractal levels

Fractal Level		Grid		Active Power Appliances
	Wire	Transformers	Reactive Power Devices	Appliances that Mainly Produce, Consume or Store Active Power
Level 1. → HV_ElA	Very long	Very large	Very large	Very large
Level 2. → MV_ElA	Long	Large	Large	Large
Level 3. → LV_ElA	Short	Medium	Medium	Medium
Level 4. → CP_ElA	Very short	Small?	Small	Small
Level 5. → Dev_ElA	Tiny	Tiny	Tiny	Tiny

High voltage

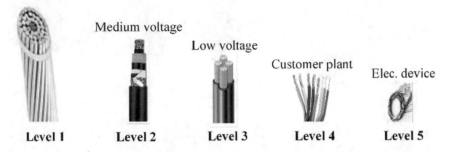

Level 1　　　　　Level 2　　　　　Level 3　　　　　Level 4　　　　　Level 5

Fig. 2.3 Wires in different levels of Smart Grids

media. Power system appliances are categorised into two groups: The grid itself and Active Power Appliances (APAs).

The grid is a construction providing the connections amongst the points of power production and consumption. The relevant grid elements are wires, transformers, and Reactive Power Devices (RPDs) [8]. It is a fact that power, whether using alternating or direct current, needs wires to be conducted from production to consumption. Thus, conductors are present everywhere in Smart Grids, Fig. 2.3.

In level 1, the HV level, wires are called lines, while in the MV and LV levels, i.e. levels 2 and 3, they are more commonly referred to as feeders. HV lines are the longest, ranging from 100 to 2000 km. The world's longest power transmission line is the 600 kV direct current Rio Madeira transmission link in Brazil [21], with a length of 2385 km. Feeders in MV grids are long, ranging from a few km to 70 km, while in LV grids, they are short and range between several tens of meters and two kilometres: characteristic for the European distributed grids [6].

CPs are included in level 4. They did not fall within the scope of power systems, and therefore their grid was never analysed. The LV grid extends within the CPs and typically maintains the radial structure. Figure 2.4 shows the electricity grid in the customer plant's environment. Various sockets and electrical devices are connected to the LV-CP connection node via wires installed in building walls. The cables are short. They range between a few to several tens of meters.

Electrical devices connected to the customer plant grid have internal small/tiny wires that range from a few to several tens of centimetres. Thus, as the fractal level increases, the wire length decreases. The same trend also characterises the wire diameters. The wire length and diameter repeatedly become smaller in the branches from level 1 upwards. Thus, wires are constructs that repeat themselves at progressively smaller scales.

The use of different voltage levels in the power system design dictated transformers' introduction as connecting elements, Fig. 2.5. Very large-, large-, and medium-sized transformers are used in HV, MV, and LV (or fractal levels 1, 2, and 3), respectively. Transformers do not exist at the CP level; the grey elements in

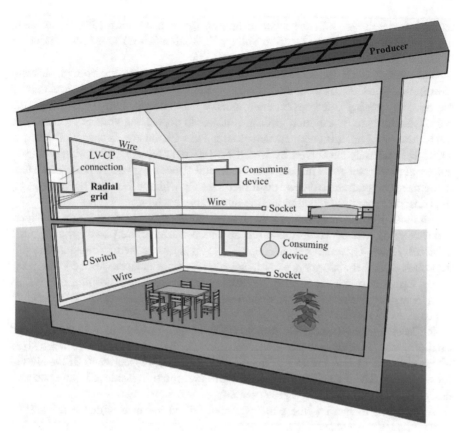

Fig. 2.4 Overview of the electricity grid in the customer's plant environment (drawing S. Lengauer)

Table 2.1. Thus, the fractal structure presented in Table 2.1 shows a design anomaly in level 4.

This anomaly may indicate further optimisation potential of the power system structure. In the past, losses were one of the key optimisation criteria in the power system design process. Only in the 1970s was it discovered that supplying a load

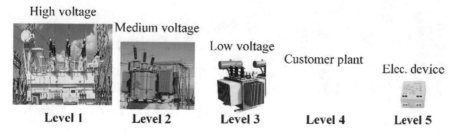

Fig. 2.5 Transformers in different fractal levels of Smart Grids

using lower voltages reduces demand and energy consumption [55]. Since then, many studies based on Conservation Voltage Reduction (CVR) have been conducted to assess the implementation's effectiveness [55].

The current CVR implementation is based on controlling the voltage through distribution grids. Apart from the fact that its implementation in broad areas carries the risk of exposing some customers to unacceptable under-voltage conditions, the radial structures in distribution grids are subject to decreasing voltage profiles. The latter means that not all customers can be supplied with the lowest permitted voltage. The last issue may be solved by installing 1:1 transformers with voltage control ability at the CP level, as indicated in the fractal structure given in Table 2.1. Their adjustment can guarantee the end customer's supply with the lowest possible voltage, thus ensuring maximum load reduction and energy savings.

Additionally, this may lead to the liberalisation of the voltage limits in super-ordinate grids, which would make the operation of the entire power system more efficient. However, fundamental economic studies are needed to substantiate this hypothesis. Tiny transformers exist in several customer devices (level 5). As in the wires' case, the transformer size decreases with voltage level, meaning that it falls with increasing fractal level. Transformers are thus constructs that repeat themselves at progressively smaller scales.

RPDs are mainly Locally Controlled (LC). They are widely used in the high and medium voltage grids to control the voltage, Fig. 2.6. Very large and large-sized RPDs are used in HV and MV (fractal levels 1 and 2), respectively. RPDs are not present at the LV level. The fractal structure presented in Table 2.1 thus shows a design anomaly in level 3, the grey element.

In contrast, in level 4, the table is already filled because with the integration of rooftop PVs, associated inverters are used to support the voltage control in low voltage grids by providing reactive power (almost purely inductive) [3]. Studies have shown that using customers' devices to control the voltage on low voltage grids is technically ineffective [28, 38] and can lead to social problems [61]. The use of the DSOs' reactive power sinks to control LV grids' voltage is found to be more effective [38]. Thus, level 3 of the fractal structure Table can be filled by DSOs, and customers'

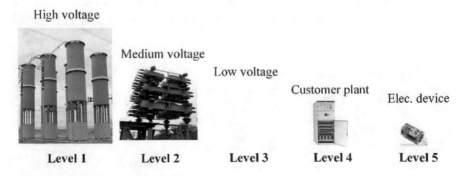

Fig. 2.6 Reactive devices in different levels of Smart Grids

| Rotating mashine | PV cells | Battericn |

Fig. 2.7 Various active power devices

inverters should be used only to compensate for their reactive power requirements [28]. Tiny RPDs exist in several customer devices (level 5). The RPD size follows the same trend as discussed above; it decreases with increasing fractal levels. RPDs are constructs that repeat themselves at progressively smaller scales.

APAs are appliances that mainly produce, consume, or store active power, i.e., rotating machines, photovoltaic, batteries, etc., Fig. 2.7. Very large-sized APAs present in hydro, nuclear, etc., power plants, and pumped hydroelectric storages are connected to HV grids (level 1). With the emergence of distributed energy resources, large, medium, and small-sized APAs are already connected in MV and LV grids and CPs (levels 2, 3, and 4), respectively. Moreover, several devices, such as computers, have tiny APAs in the form of batteries (level 5). The column of APAs in Table 2.1 is filled. The APA size follows the same trend as discussed above; it decreases with increasing fractal levels. Thus, APAs are constructs that repeat themselves at progressively smaller scales.

As a result, although very heterogeneous, the construction of Smart Grids is self-similar. Figure 2.8 shows its fractal pattern. Both groups of self-similar constructs (Grid and APA) are figuratively wrapped in an ellipse, Fig. 2.8a.

> By definition, the **fractal pattern** of Smart Grids consists of Electrical Appliances (*ElA*) designed for a pre-defined level.

Each *ElA*-ellipse denotes a separate chain link. It includes the grid of a specific level and all APAs connected to it. Figure 2.8b illustrates *ElA* in different levels: HV-*ElA*, MV-*ElA*, LV-*ElA*, CP-*ElA* and Dev-*ElA* for levels 1, 2, 3, 4 and 5, respectively. Figure 2.8.c magnifies the CP- and Dev-*ElA* patterns.

2.3.3 Fractal Set of Smart Grids

The discovering of the Smart Grid's fractal set needs the investigation of its holistic composition. A Smart Grid is the giant electromagnetic machine of our time. It

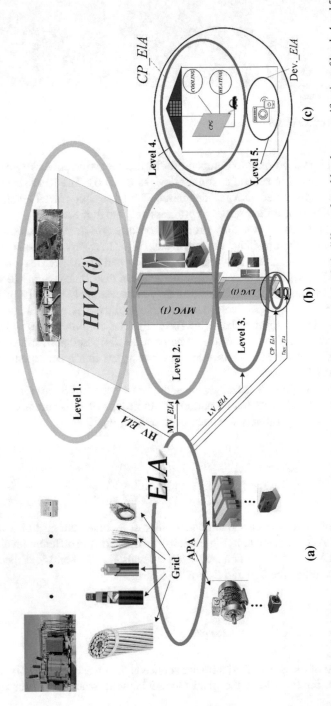

Fig. 2.8 Link chain fractal of power systems: **a** The fractal pattern electrical appliances (*ElA*); **b** *ElA* in different fractal levels; **c** magnification of levels 4 and 5

stretches across continental space and is owned, operated, and used by various entities. Given the different fractal levels discussed in the fractal pattern identification (Sect. 2.3.2), the Smart Grid is divided into four typical zones. Figure 2.9 shows all four zones as HV-, MV-, LV- and CP zone. The electrical device zone is not established because they are wastefully small. MV-zones are connected to the corresponding HV-zone via a supplying transformer (included in a supplying substation). LV-zones are connected to the corresponding MV-zones via a distribution transformer (included in a distribution substation). Consumers use CP-zones. The latter are usually directly connected to the corresponding LV-zones.

From the bird's-eye view, all identified fractal levels are in the same territory, T. The overall pattern of level 1, HV-*ElA*, spreads over the entire T of a country or a part of it, Fig. 2.10. The other fractal levels' patterns propagate in smaller areas of the same territory T, repeating themselves many times. The repeating number of patterns in different fractal levels is derived from the relationship between different zones of power systems in real conditions and the number of appliances comprised in each of them, Table 2.2.

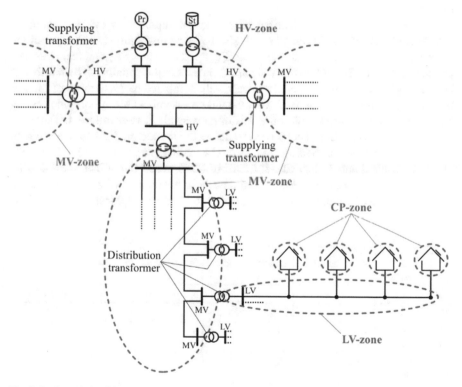

Fig. 2.9 Overview of the power system zones

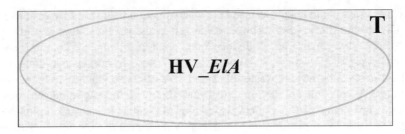

Fig. 2.10 The fractal set at HV level

Since the MV and LV grids have radial structures, the number of MV and LV zones is defined by the number of suppling (HV/MV) and distribution substations (MV/LV).

Table 2.2 shows a structural overview of the power system in real conditions and of its fractal. A European power system, operated by one TSO, has about 300 supplying and 100,000 distribution substations to supply about 4 million customer plants.

It is assumed that each distribution substation supplies on average about 40 customer plants [36]. Additionally, it is also believed that there is an average of ten electrical devices in each customer plant. Figure 2.11a shows a graphical overview of the structure of Smart Grids. The graph increases exponentially as the number of customer plants, 4 million, increases drastically compared to the number of the HV zones, that is one. Thus, the fractal set cannot be assembled using only one scaling factor because the number of zones in different levels is very different. Table 2.2 shows the various scaling factors used to define the number of *ElAs* for each fractal level and the derived numbers of *ElAs* used. The scaling factor for level 1 is set to one, while for levels 2 and 3, they are set to 50 and 5555, respectively. The same scaling factor of 55,555 is used for levels 4 and 5 because the number of zones included in them differs only by a factor of ten. Figure 2.11b shows a graphical overview of the

Table 2.2 The fractal structure of Smart Grids based on different fractal levels

Entire real power system	No. of zones	Scaling factors	Fractal of the entire power system	
Zone			No. of ElAs	Fractal Level
HV	1	1	1	Level 1HV
MV (Suppl. Substations)	300	50	6	Level 2MV
LV (Distr. Substations)	100,000	5555	18	Level 3LV
CP (Customers)	4,000,000	55,555	72	Level 4CP
Dev. (Elec. Devices)	40,000,000	55,555	720	Level 5Dev

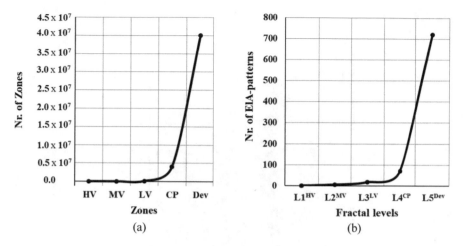

Fig. 2.11 Overview of the structure of Smart Grids: **a** Real conditions; **b** Fractal

structure of the fractal set. The graph behaves similarly to one of the power systems in real conditions.

Using this approach, each of the six MV-*ElA*s touches the single HV-*ElA* at one point and tangentially to each other, Fig. 2.12. The latter symbolizes the tie switches (normally open points) connecting feeders of different MV-subsystems. For the 18 LV-*ElA*s is used the same principle as for MV-*ElA*s, Fig. 2.13. Four CP-*ElA*s are added into every LV-*ElA*, each touching only one point of the LV-*ElA* pattern. The touching point reflects that each customer plant has an electrical connection point to the LV network and no connection to each other, Fig. 2.14.

The same principle is used to add the ten Dev-*ElA*s into every CP-*ElA*, Fig. 2.15. The fractal pattern *ElA* of different levels is assembled in a fractal set; a link chain fractal limited to five interlinks.

The calculation of the fractal dimension verifies the significance of the fractal set of Smart Grids. It contains information about the geometrical structure at multiple scales reflecting the fractal set's complexity: The greater the complexity, the larger the fractal dimension.

2.3.4 Fractal Dimension

Fractals can be characterised by the fractal dimension D. For a fractal located in a bi-dimensional space, D is larger than one and less than two,

$$1 < D < 2. \tag{2.1}$$

The larger D, the greater is the fractal complexity.

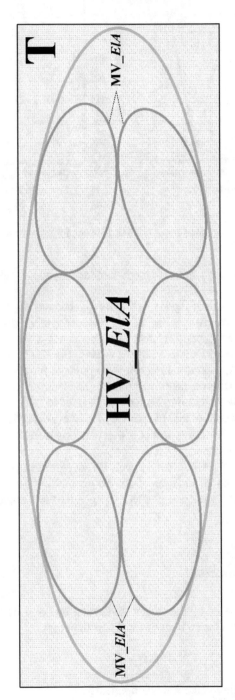

Fig. 2.12 The fractal set including HV and MV levels

Fig. 2.13 The fractal set including HV, MV and LV levels

Fig. 2.14 The fractal set including HV, MV, LV and CP levels

Fig. 2.15 The fractal set of Smart Grids (HV, MV, LV, CP and Dev. Levels)

D is calculated using the box-counting method, which is based on counting boxes with side length (s) occupied by the fractal. The number of occupied boxes, N, increases by making the grid finer (s → 0). D is estimated as the exponent of a power low as follows:

$$D = \lim_{s \to 0} \frac{\log N(s)}{\log\left(\frac{1}{s}\right)} \qquad (2.2)$$

The software Fractalyse 2.4 [19] is used to calculate the fractal dimension of the entire power system, customer plants, and electrical devices. The fractal dimension is obtained by using a logarithmic, linear regression to fulfil the following objective function

$$\log N(s) = D \cdot \log\left(\frac{1}{s}\right) + C, \qquad (2.3)$$

where C is the equation constant.

Figure 2.16 shows the separate fractal sets of Smart Grids at different levels, while the logarithm diagram for measuring D of each level is shown in Fig. 2.17. The segments' slope increases steadily with increasing fractal level. Figure 2.18 shows the fractal dimension D for different separate fractal levels. D increases from 1184 for level one (HV) to 1272 and 1308 for levels two and three (MV and LV), respectively. The D-increase from level one to level two is significant ($\Delta D^{L1 \to L2} = 0.088$), which means that the complexity of Smart Grids at the MV level increases dramatically compared to the HV level. Another significant D-increase, even bigger than the previous one, is observed between levels three (LV) and four (CP) ($\Delta D^{L3 \to L4} = 0.108$). The complexity of Smart Grids at the CP level increases considerably compared to the LV level. The slope of the trend line of levels four and five (dashed line) is more significant than in the trend line for levels two and three (dotted line), 0.156 and 0.0347, respectively.

Compared to the customer plant, the increase in complexity at the electrical device level is more substantial than that between the LV and MV levels.

Usually, the resources required to develop and purchase the different parts of a complex system are proportional to their complexity. The fractal dimension analysis indicates that Smart Grids' realisation requires the highest global resources to develop and purchase electrical devices, followed by a continuous reduction for CPs, LV, and MV up to the lowest resource allocation for the HV level.

In conclusion, fractal analysis is a suitable instrument for characterizing the Smart Grid structure's heterogeneity, as it can represent the spatial differences within the Smart Grid. The identified fractal pattern *EIA* consists of unique and independent

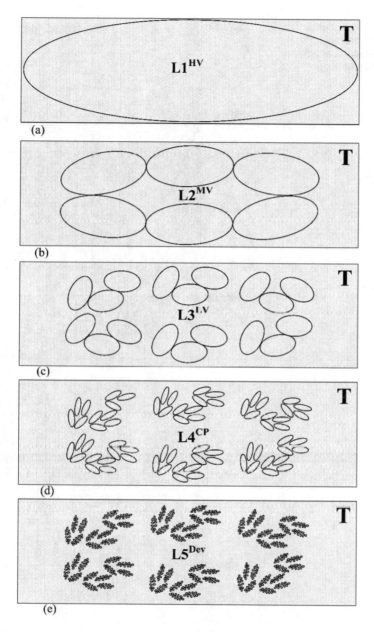

Fig. 2.16 Smart Grid's separate fractal sets in different levels: **a** Level 1; **b** level 2; **c** level 3; **d** level 4; **e** level 5

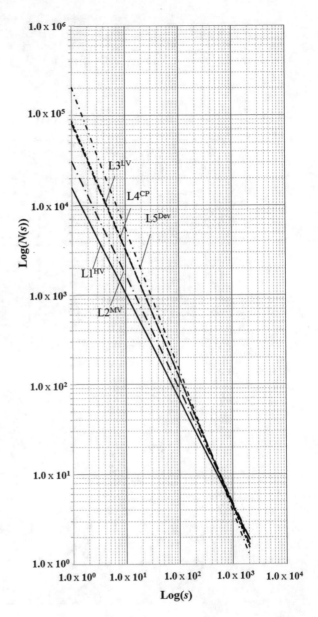

Fig. 2.17 Logarithmic diagram for measuring the fractal dimension of different separate fractal levels

elements, thus fulfilling the modern architecting theory's basic requirements. *EIA* establishes the fundament of the Smart Grid paradigm.

Fig. 2.18 The fractal dimension D for different separate fractal levels

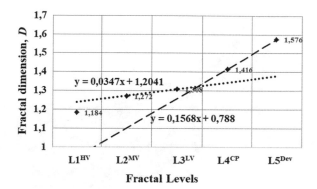

2.4 *LINK*-Paradigm and the Associated Holistic Model

The realisation of Smart Grids is related to a growing demand for more sensors, more communication, more computation, and more control [4], these represent an ever-growing cornucopia of new technologies [49]. Therefore, the *ElA* fractal pattern is combined with control schemes and interfaces to create the Smart Grid paradigm Fig. 2.19.

> By definition, the *LINK*-**Paradigm** is a set of one or more *ElA*s, i.e., a grid part, storage device, or a producer device, the controlling schema, and the interface.

The *LINK*-paradigm is used as an instrument to design a *LINK*-based holistic architecture. It facilitates modelling of the entire power system from high to low voltage levels, including CPs. It includes the description of all power system operation processes such as load-generation balance, voltage assessment, dynamic security, price and emergency driven demand response, etc. [69]. The *LINK*-Paradigm is fundamental to the holistic, technical, and market-related Smart Grid model with large DER shares. Figure 2.20a shows the holistic technical model (the "Energy Supply Chain Net"), which extends to fractal level 4. It illustrates the links' compositions and their relative position in space, both horizontally and vertically. In the

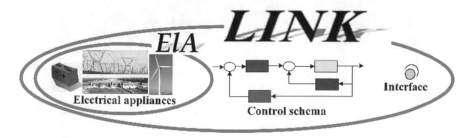

Fig. 2.19 Overview of the Smart Grid paradigm *LINK*

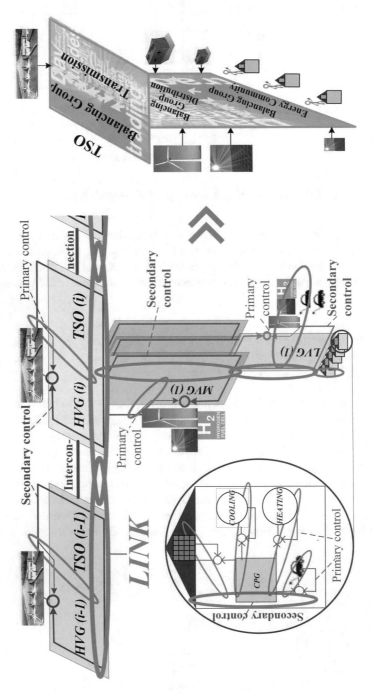

Fig. 2.20 Overview of the holistic models: **a** Zoom in CP; **b** Technical model the "Energy supply chain net"; **c** Market model

horizontal axis, the interconnected High Voltage Grids (HVG) are arranged. They are owned and operated by Transmission System Operators (TSO). Medium (MVG) and Low Voltage Grids (LVG) and the Customer Plant Grids (CPG), including the HVG to which the MVG is connected, are set vertically. MVGs and LVGs are owned and operated by the DSOs, while customers use CPs. Electricity producers (hydro-electric power plants, wind and PV plants, etc.) and storage (pumped hydroelectric power plants, batteries, e-cars bateries, in the form of heating and cooling, hydrogen production, etc.) are connected at all levels.

By definition, an "**Energy Supply Chain Net**" is a set of automated power grids intended for chain links (abbreviated as links), which fit into one another to establish a flexible and reliable electrical connection. Each link or link bundle operates autonomously and has contractual arrangements with other relevant boundary links or link bundles [26].

The holistic model associated with the energy market is derived from the holistic technical model, the "Energy Supply Chain Net," as shown in Fig. 2.20b. The whole energy market consists of coupled market areas (balancing groups) at the horizontal and vertical axes. TSOs operate on the horizontal axis of the holistic market model, while DSOs operate on the vertical. Based on this model, TSOs and DSOs will communicate directly with the whole market to ensure a congestion-free distribution grid operation and take over the task of load-production balance. The owner of the distributed energy resources as well as the prosumers (producers and consumers of electricity) may participate directly in the market or may do so via aggregators or Energy Communities (ECs) [35]. The Local Retail Markets (LRM) creation attracts the Demand Response (DR) bids and stimulates investment in the Energy Communities areas (see Sects. 2.5.7 and 3.3).

2.5 *LINK*-Based Holistic Architecture

The electricity supply structure changes radically because of many DG units, each with the possibility of interfering with the system operation at all voltage levels. The utilization of the added flexibility necessitates a new architecture that should guarantee to perform as expected by unifying their whole structure and systematizing operational tasks. It should handle the scalability of the system and minimize the complexity.

By definition: A **holistic power system architecture** is an architecture where all relevant components of the power system are merged into one single structure. These components could comprise of the following:

- Electricity producer (regardless of technology or size, e.g., big power plants, distributed generations, etc.);
- Electricity storage (regardless of technology or size, e.g., pumped power plants, batteries, etc.);
- Electricity grid (regardless of voltage level, e.g., high-, medium- and low-voltage grid);
- Customer plants; and
- Electricity market

The holistic architecture unifies all interactions within the power system itself, between the network-, generation- and storage operators, consumers and prosumers, and the market, thus creating the possibility to harmonize them without compromising data privacy and cybersecurity. It facilitates all necessary processes for a reliable, economical, and environmentally friendly operation of smart power systems. It allows a clear description of the relationships between different actors. It creates conditions to go through the transition phase without causing problems. It enables Energy Communities, the Sector Coupling and coupling with the non-energy sectors such as health and comfort [32]).

2.5.1   Transformation of Power Industry Structure

Historically, the power industry is perceived as an assembly of three main components: Electricity producers included in power plants, Grid, and Customers. Electricity storage, almost the pumped hydroelectric plants, is traditionally presented as part of the power plant's component. While the grid is split into two main parts:

- The transmission grid comprising very high and high voltage grids, and.
- The distribution grid containing medium and low voltage grids.

Utilities are vertically integrated and own and control electricity producers, transmission, and distribution components. Customers consume the electricity offered by the nearest electrical utility. Figure 2.21 illustrates the structural transformation of the power industry and the electricity customers. The traditional structure of power systems is shown in Fig. 2.21a.

The decentralization process of the electricity industry was in progress in the 1990s. At that time, various legal entities such as generation, transmission and distribution companies, market operators, political decision-makers, etc., were emerged from the vertically integrated utilities. Storage, almost pumped hydroelectric plants, were treated as a generation company.

The last 30 years have been characterised by an extraordinary development of various technologies that significantly impact the power industry structure. These developments change the perception of the components of the power supply system. In addition to dividing the vertically integrated utilities into different legal entities, two other fundamental issues are affecting its structure:

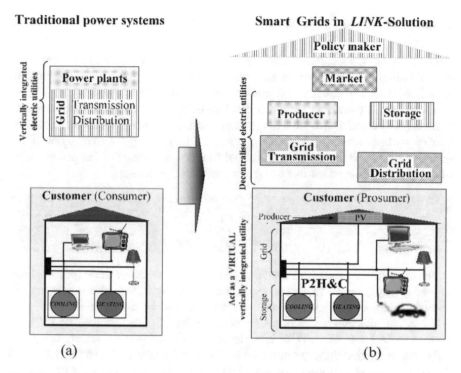

Fig. 2.21 Overview of the structural transformation of the power industry and electricity customers: **a** Traditional power systems; **b** Smart Grids and Sector Coupling in the *LINK*-Solution

1. Nowadays, storages are undergoing an intensive development process. Diverse technologies are developed. They are available in different sizes and can be integrated into any voltage level of the grid. The electricity power surpluses might be stored or used in other sectors, Power-to-X (P2X). That means that from the perspective of the power system, Sector Coupling is a storage process. Treating storage as part of a power plant seems to be no longer appropriate: It cannot be used to describe P2X processes. Storage might be perceived as one of the new architecture's main components.

2. Customers are experiencing radical changes. They are not only the owner of devices that consume electricity, but they also own a grid (see Sect. 2.3.2, Fig. 2.4), electricity producers such as PV facilities, and storage devices such as batteries. Additionally, the customer can store the renewable Power surplus on Heating and Cooling devices, Power-to-Heating and Cooling (P2H&C) or Power-to-Thermal (P2T). Based on these facts and the vertically integrated utility definition, prosumers might be perceived as virtual vertically integrated utilities [27], Fig. 2.21b.

2.5.2   Architecture Components and Structure

Architectural components are unique parts that together form the architectural structure. Fleshing out the holistic model into a schematic holistic architecture requires specification of the principal independent architecture components.

Electricity is an energy form that can be easily transported and distributed, but there are no possibilities to store it. For this reason, other energy forms with available carriers are used for its production and storage. Figure 2.22 gives an overview of the energy carriers and means of transport used in the energy conversion process. One of the elements needed for this process is coupling components.

Coupling components are devices that serve to convert energy into other forms.

In nature, there are various available energy carriers such as water (potential and kinetic energy); wind (kinetic energy); the ray of sunlight (radiant energy); gas, oil, and coal (chemical energy); radioactive elements (nuclear energy) and so on.

Coupling elements such as generators (combined with turbines) or PV cells convert these kinds of energies into electricity. The coupling elements are popularly known as electricity producers. Electricity transports and distributes through the grid to the end-customers to converse in other energy forms that can not be used anymore (consumption). Other coupling components, such as pumps, fuel cells, anode and cathodes (for batteries), heaters, and so on, convert the electricity surplus into other energy forms to be stored and reconverted in electricity at another point in time (bi-directional energy conversion) or to be consumed further. These coupling components are known as electricity storage units. Pumped hydroelectric power plant is the classical electricity storage case using water as an energy carrier: electricity converts into potential energy. Water may also be used as a thermal energy carrier. In this case, the water temperature increases using the electricity surplus. With a coupling component between the warm and cold water, a consumable water temperature is reached.

Based on the analyses before, three fundamental components of the architecture are identified: Producer, Storage, and Grid. They are an integral part of one of the three constituent elements of the *LINK*-Paradigm: Electrical appliances. The components of the holistic architecture are Producer-Link, Storage-Link, and Grid-Link.

2.5.2.1   Producer-Link

The first fundamental component of the holistic architecture is the Producer-Link. Figure 2.23 shows an overview of it.

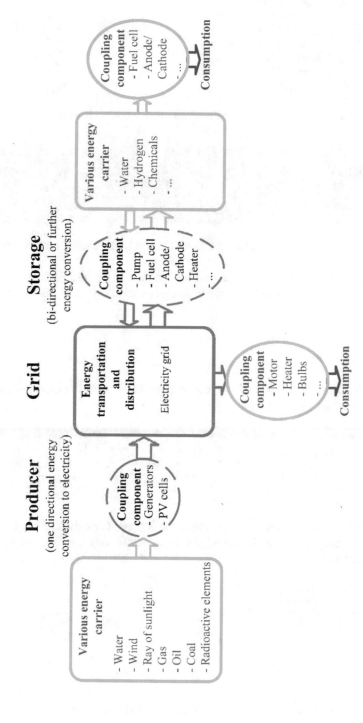

Fig. 2.22   Overview of the energy carriers and means of transport used in the energy conversion process

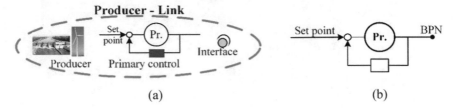

(a)   (b)

Fig. 2.23 Overview of the Producer-Link: **a** General depiction; **b** Control scheme

By definition, **Producer-Link** is a composition of an electricity production facility, be it a generator, photovoltaic, etc., its Primary Control (PC), and the interface, Fig. 2.23a.

Primary control refers to closed-loop control actions taken locally (device level) (see A.2.2.2).

Figure 2.23b shows the electrical scheme of the Producer-Link. Each Producer has a Boundary Producer Node (BPN) via which it connects with the Link-Grid and feeds the electricity.

2.5.2.2 Storage-Link

The second fundamental component of the new architecture is the Storage-Link. Figure 2.24 shows an overview of it.

By definition, **Storage-Link** is a composition of a storage facility, be it the generator of a pumped hydroelectric storage, batteries, etc., its PC, and the interface, Fig. 2.24a.

Figure 2.24b shows the electrical scheme of the Storage-Link. Each Storage has one Boundary Storage Node (BSN) via which it connects to the Link-Grid and feeds or consumes electricity.

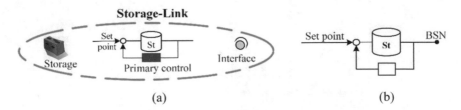

(a)   (b)

Fig. 2.24 Overview of the Storage-Link: **a** General depiction; **b** Control scheme

2.5.2.3 Grid-Link

The third and most complex fundamental component of the new architecture is the Grid-Link. Figure 2.25a shows an overview of it, while Fig. 2.25b illustrates the Link-Grid.

> By definition, **Grid-Link** is a composition of a grid part, called Link-Grid, with the corresponding Secondary Control (SC) and the interface, Fig. 2.25.

> By definition, **Link-Grid** is the grid part included within the Link. It refers to electrical equipment like lines/cables, transformers, and reactive power devices connected directly to each other by forming an electrical unity. Link-Grid size is variable and is defined from the area where the Secondary-Control is set up.

Thus the Link-Grid may include, e.g., one subsystem (the supplying transformer and the feeders supplied from it) or a part of the sub-transmission network, as long as the SC is set up on the respective area. As a result, depending on its size, the Link may represent the high-, medium-, low- and even the customer plant grid.

Figure 2.26 shows the control schemes set on a typical Link-Grid. Figure 2.26a depicts the Hertz/Watt secondary control, while Fig. 2.26b the Volt/var secondary control. Each Link-Grid has many Boundary Link Nodes (BLiN) through which it connects with neighbouring Link-Grids. The neighbouring Link-Grids are represented with the symbol #. Producers inject directly into it via BPN; Storages inject or consume power via BSN.

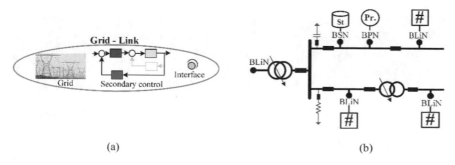

(a)   (b)

Fig. 2.25 Overview of the Grid-Link: **a** General depiction; **b** Link-Grid

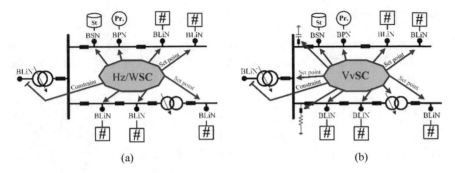

Fig. 2.26 The control schemes set on a typical Link-Grid: **a** Hertz/Watt secondary control; **b** Volt/var secondary control

> By definition, the **Link-Grid** is upgraded with secondary control for both significant power systems entities pairs the frequency/active power and voltage/reactive power based on the fact that the frequency depends on active power whereas voltage mainly on reactive power.

Its algorithm needs to fulfil technical issues and calculate the connected facilities' primary controls' setpoints by respecting the dynamic constraints necessary to enable a stable operation (see Sects. 2.6.2 and 2.6.3). The Link-Grid's facilities, such as transformers and the reactive power devices, are almost upgraded with primary or local control. Thus, SC sends setpoints to its facilities and all entities connected at the boundary nodes.

2.5.2.4 Structure

Data privacy and significant data transfer are the two biggest challenges that Smart Grids technologies are facing today. The distributed *LINK*-based architecture meets these two challenges [22], Fig. 2.27.

> The *LINK*-based architecture's fundamental principle is to forbid access to all resources by default, allowing access only through well-defined boundary points via interfaces.

Each Link act as a black-box that exchanges a predefined minimum data set with the other Links. The different links communicate via well-defined technical interfaces "T" (see Sect. 2.5.5.4).

The new structure shown in the Figure has two parts: the power system and customer plant parts. Since the culmination of the decentralization process, the power

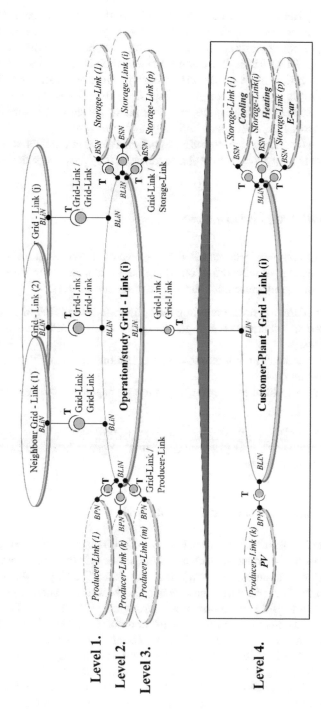

Fig. 2.27 Overview of the architecture structure

system consists of different companies. They have to communicate via external interfaces that are subject to data privacy and cybersecurity. Since all three architecture components are present at all three fractal levels, they are presented one above the other in the Figure. The Grid is the central element that connects Producers and Storage across the entire Smart Grids. Producer- and Storage-Links communicate with the Grid-Links via interfaces. Neighbouring Grid-Links also communicate with one another via interfaces. In this Figure, customer plants are shown separately from the rest of the Smart Grids, acting as a vertically integrated entity (see Sect. 2.5.1). Different Link types available in the same customer plant have internal interfaces that are neither subject to data privacy nor cybersecurity, as they have the same owner.

2.5.3   Different Architecture Levels

The architecture of complex systems usually has several architectural levels representing different degrees of abstraction on which the system can be modelled. The holistic *LINK*-based architecture has four basic architectural levels:

- The holistic architectural level including the entire Smart Grids and the market;
- The generalized architectural level representing the entire Smart Grids in a very compact manner;
- The technical/functional architectural level representing Smart Grids in all their technical levels; and,
- The physical architectural level referring to the management systems of Smart Grids.

To better understand the various architectural levels, the technical/functional level is addressed first, followed by the physical one. Then the generalised and the holistic architectural levels are discussed.

2.5.3.1   Technical/Functional Architectural Level

The technical/functional architectural level includes the conjunction of all three architecture elements such as Producer-, Grid- and Storage-Links in the four-fractal levels high-, medium-, low- and customer plant levels, Fig. 2.28. The High-Voltage_Grid-Link is set up on the HV grid part. The Producer-Links and Storage-Links electrically connected to this grid part communicate through the T interfaces. The neighbouring Grid-Links, be high or medium voltage, communicate via "T" interfaces as well. The Medium- and Low-Voltage_Grid-Link are set up on medium and low grid parts, respectively. The Producer-Links and Storage-Links correspondingly electrically connected to these grid parts communicate through the "T" interfaces. The neighbouring Grid-Links, be high and low voltage, or medium voltage and customer plants, also communicate via "T" interfaces. In customer plant level repeats the same

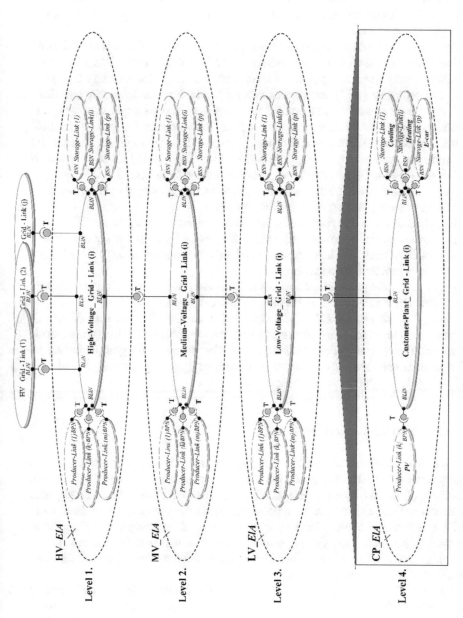

Fig. 2.28 Technical/functional architectural level of *LINK*-based architecture

structure. The Customer-Plant_Grid-Link is set up on the CP grid. The Producer-Links, e.g., rooftop PV-facilities, and Storage-Links, e.g., the battery of an e-car electrically connected to this grid, communicate through the "T" interfaces.

LV_Grid-Link is the only neighbour of a CP_Grid-Link, which also communicates through "T" interfaces.

Practical Grid-Links areas do not have to be limited to the traditional categorisation of the power grid, such as high-, medium- and low voltage levels. The area size is variable and may contain only a particular part of the corresponding voltage level grid, or it may comprise grids from various voltage levels. For example, only one Grid-Link may be set up over an MV and LV grid if the applications required to set up the secondary control on this grid part are available. This architectural level is detailed, facilitating all technical/functional processes required for Smart Grids' reliable and economical operation. The Unified Modelling Language (UML) diagrams are used to describe all Smart Grids operation processes (example in Sect. 2.6.5).

Standardised structures characterise the technical/functional architecture level that comprises all technical levels of Smart Grids.

It enables the step into the Physical and Generalised architectural levels.

Due to the vast size of Smart Grids and, consequently, the large amount of technical data required to optimise their operation and planning, a calculation of Smart Grids as a whole is practically impossible.

Hence, the **main principle of the *LINK*-Solution** is the optimisation of the whole Smart Grids by coordinating and adapting the locally optimised Links.

Figure 2.29 shows the *LINK*-Technology that consists of a network of coordinated primary and secondary controls to ensure a feasible and resilient Smart Grids behaviour. The secondary control units set on different parts of the grid (including the CP grid) communicate and coordinate by adjusting all primary controls of producers and storage belonging to their operation area.

2.5.3.2 Physical Architectural Level

The physical architectural level is located at the lowest level of abstraction, so it is highly detail-oriented and refers to Smart Grids' management systems. It deals with specific applications, data representations, and other technical issues. The physical architectural level aims to enable a particular technology solution to be realised in real life. It serves as a guide for the engineering team developing and implementing the system. As such, architecture elements refer to actual real applications, server modules, operational processes, etc. Therefore, detailed modelling is perhaps more important at this level than anywhere else.

Figure 2.30 shows an overview of the physical architectural level used in project ZUQDE to realise the Volt/var control process in a closed-loop [41]. The secondary control is realised using two fundamental applications: the Distribution system State Estimator (DSE) and Volt/var Secondary Control (VvSC). The applied algorithm calculated the set points by respecting the constraint set to the HV/MV transformers through a constant $\cos\varphi$. The setpoints, reactive power Q and voltage, were sent

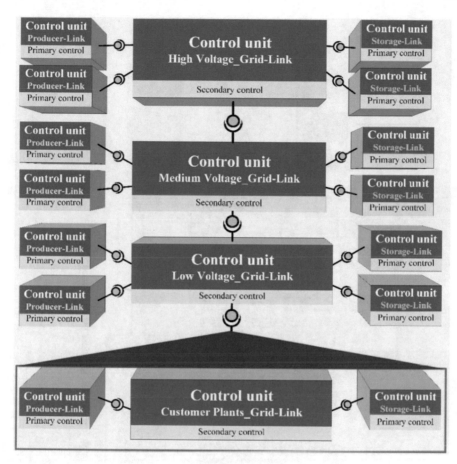

Fig. 2.29 *LINK*-Technology: Standardised control structure over the entire power system and customer plants

to all four "run of river" DGs and the feeder's head bus bar, respectively. All relevant generators were upgraded along with the primary control, thus building up the Producer-Link component. All distributed transformers and the corresponding LV grids were modelled as loads.

2.5.3.3 Generalised Architectural Level

The generalised level is the most abstract one of the three. It is a very high position from which all or most things can be seen. It has the disadvantage that the details are not visible. The generalised architecture layer effectively acts as a structural model to highlight essential appliances' relationships but not how they work. This means

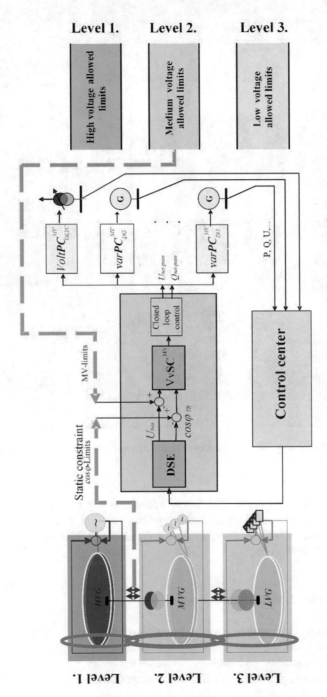

Fig. 2.30 Overview of the physical architectural level used in project ZUQDE to realise the Volt/var control

that the point here is not to reveal the process flow but to establish relationships with other systems on the same abstraction level.

Thanks to fractal structures, Smart Grids can be presented very compactly, as shown in Fig. 2.31. The technical/functional architecture's standardised design allows the transition to the generalised architecture level, where the four fractal levels are represented very compactly.

2.5.3.4   Holistic Architectural Level

The holistic architecture level is the core of the *LINK*-based architecture Fig. 2.32. It comprises the Smart Grids and the electricity market. The market surrounds the generalised architecture and communicates with it through the market interfaces "M" by exchanging aggregated meter readings, external schedules, etc. [33]. At this architectural level, the grid links of customer plants are removed from the generalised presentation because they are too small to participate directly in the whole market. They may participate in the common market through aggregators or energy communities [35]. For the sake of privacy and cybersecurity, the market interfaces "M" are designed apart from technical interfaces "T".

This architectural level forms the base for designing and implementing the demand response process, the most comprehensive and complex operation process in Smart Grids: All voltage levels, customers, and the market are affected here (see Sect. 2.6.5).

2.5.4   LINK Operation

The main goal of Link operation is to run it in a safe, reliable, economical, and environment-friendly manner. The Link-Operator is one of the three main actors of the Smart Grid operation [69], leading all required processes to deliver electricity to all economic's sectors and private consumers that adhere to strict tolerances for voltage and frequency. He manages the Link or Link-bundle using a multi-computer system within a control centre. The Link-Operator interacts with the utility's field crews, technical and general staff, and the neighbouring utilities' homologs. He carefully monitors the Link behaviour using Supervisory Control and Data Acquisition (SCADA) and State Estimator (SE), if available, to anticipate potentially dangerous and costly events or mitigate their effects.

2.5.4.1   Operators

A Link-Operator is a generic term for diverse specializations. The architecture components Producer, Storage, and Grid define their domain (Fig. 2.33).

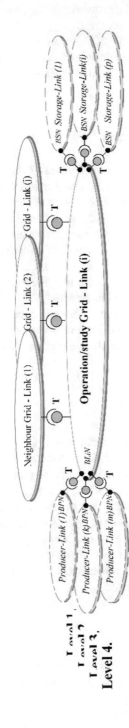

Fig. 2.31 Generalised architectural level of the *LINK*-based architecture

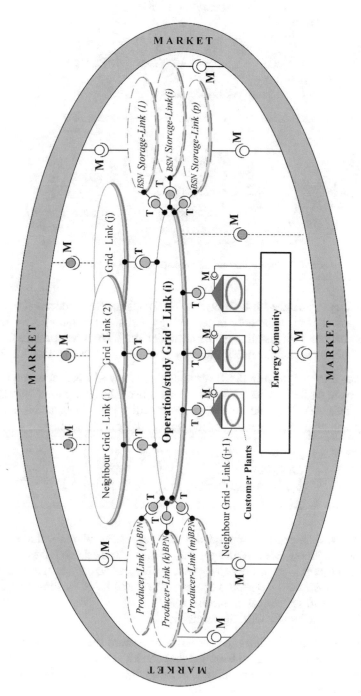

Fig. 2.32 Holistic architectural level of the *LINK*-based architecture

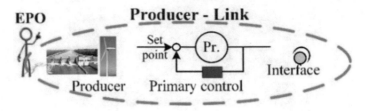

Fig. 2.33 Overview of the Producer-Link and the corresponding operator, EPO

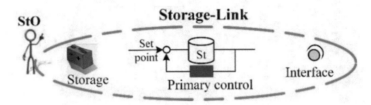

Fig. 2.34 Overview of the Producer-Link and the corresponding operator, StO

1. The **Electricity Producer-Link Operator** (EPO) operates the Producer-Links set up on every power plant regardless of technology and size (excluding very small power plants, such as PVs installed on the customer side). He is responsible for the operational planning and maintaining the power generation schedule, Fig. 2.34).

2. The **Storage-Link Operator** (StO) operates the Storage-Links set up on every storage facility regardless of technology and size (excluding very small storages, for example, batteries, installed on the customer side). He is responsible for the operational planning and maintaining of the storage schedule, Fig. 2.34.

> By definition, the **Storage Operator** is a natural or legal person that provides and stores electric energy to or from other natural or legal persons.

3. The **Grid-Link Operators** operate the grid regardless of the voltage level. Figure 2.35 shows an overview of the vertical chain of Grid-Links. Nowadays, the TSO operates the HVG, so in the *LINK*-Architecture, he takes over the operation of HV_Grid-Link. Similarly, the DSO operates MVG and LVG. Customers use the CP_Grid-Link and all appliances connected to it.

Each Link or Link-bundle Grid-Operator should:

- Balance the load and the injection in real-time, where the load represents the summation of the system native load and the scheduled exchange to other Links. The injection represents the generation's summary, injection from storage devices, and the scheduled exchange to other Links;

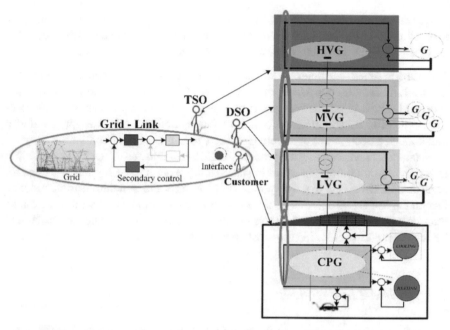

Fig. 2.35 Overview of the vertical chain of Grid-Links and the corresponding operators, TSO, DSO, and Customers

- Keep the voltage in its Grid-Link or the Grid-Link-bundle within limits by controlling the reactive power flows with the neighbouring external Links (see 2.6.2.1).
- Actively manage its Grid-Link or the Grid-Link-bundle;
- Monitor its Link-grid or the bundle of Link-grids;
- Access all the data of the Grid-Link;
- Exchange the data with the neighbouring external Grid-Links and all devices connected directly to the own Link-grid or the bundle of Link-grids;
- Have the right to use and offer services to the neighbours;
- Have the right to dispute with the neighbours to guarantee a reliable and stable operation of his Link-Grid or Link-Grid's bundle;
- Decide the actions should be taken for a secure and optimal operation of the own Grid-Link or Grid-Link-bundle;
- Be incentivised to invest in adequate solutions, beyond physical reinforcements, to increase the flexibility of the Grid-Link or Grid-Link-bundle; and.

Facilitate effective and well-functioning retail markets.

2.5.4.2 Operating Modes

LINK-Architecture facilitates two normal operating modes:

1. **Autonomous** – Each individual Link or Link-bundle operates independently by respecting the contractual arrangements with other relevant boundary Links or Link-bundles. All Links are connected, creating an extensive power system.

2. **Autarkic or self-sufficient** – It is an optional normal operating mode that applies in any Link-bundle, consisting of at least one Grid-Link and one Producer-Link Storage-Link. It is self-sufficient andsustainable without any dependency on electricity imports. Restorationis an option of this operating mode that applies after a blackout (see Sect. 2.6.6) to supply electricity to a minimum of appliances.

A familiar resynchronisation process should be established to switch the operating mode from autonomous to autarkic successfully. Each Grid-Link has a secondary control on active and reactive power that supports the synchronisation process. Depending on the properties of the Links, the resynchronisation with other Links may be automatic or manual. However; the resynchronisation philosophies should be investigated to determine the most appropriate approach.

2.5.5   Grid-Link Modelling

For many engineers, power systems have had the same architectural shape for decades. The developed assumptions and methods worked outstanding. The latest developments in the power industry and continuous changes in their structure call into question the longstanding assumptions and approaches. One of them is the concept of "Load", discussed in the following.

2.5.5.1   Detailed Definition of "Load"

At the beginning of the power systems' history, the necessary calculations were done manually. Although they weren't that large at the beginning, their detailed analyses were almost impossible. Various calculation models were created, and the concept of the load was introduced to describe the boundaries of these models. The load represented the power that was delivered from the boundary buses to the devices or the part of the grids connected there. It had the units [A], [VA], [W] and [var] and was presented in the one-line diagrams by an arrow, " → " always pointed out from the bus. Electricity has always flowed reliably in one direction. This arrow simultaneously represented the device (as hardware) and current or power flow direction (as energy). Using the same arrow symbol to indicate the device itself (or part of the network) and the current or power flowing through the device is not optimal and may lead to confusion.

Power systems continue to grow over time, and although powerful computers are used, their detailed calculation is still impossible. Additionally, since decentralisation in the power industry, various legal companies have been founded (such as TSO,

DSO, GenCo, and so on) that own and operate different parts of the same electromagnetic machine. The use of "load" in modelling the power system is becoming increasingly important and critical. The integration of distributed energy sources provokes a two-way power flow circulation in the network and makes the detailed review and specification of the term "Load" a necessity.

For the further development of the *LINK*-Solution, the definition of the term "Load" is rendered more precisely by splitting it into the hardware which consumes the power, the power consumed, or injected as follows:

- In the online diagrams, various drawing symbols may represent the elements in which the load flows depending on the boundary point's topological position. It can represent:

  - **The flow of individual devices**, such as motors, lighting, and so on: Known as "device load" (IEEE Task Force on Load Representation for Dynamic Performance 1993) (see also A.2.1). The power flows only in one flow direction: From the network into the device. The symbol (Dev.) presents the electrical device;
  - **The aggregated flow of individual devices combined with electrical appliances of a system portion** such as sub-transmission lines, distribution feeders, customer plants wiring, reactive power devices, transformers, distributed energy resources: Known as "busload." (IEEE Task Force on Load Representation for Dynamic Performance 1993) (see also A.2.1). The power flow is bi-directional (load and injection). To be correct, the term Loadinj should be used here to combine the word "load" with "inj" from the injection. A system portion or the Link-Grid are presented by | # |;
  - **The aggregated flow of all individual component devices and appliances of a power system**, such as motors, lighting, lines, transformers, and so on: Known as "system load" [25]: It has no flow direction.

  Here it is essential to specify three cases Load, Injection, and Loadinj:

- Load definition.

> A "**Load**" is the amount of current or power that ever flows through boundary points outside the considered grid area " → ".

- Injection definition.

> An **"Injection"** is the amount of current or power that ever flows through boundary points into the considered grid area " ← ".

- Loadinj definition.

> A **"Loadinj"** is the amount of current or power that flows bi-directionally through the considered grid's boundary points " ↔ ".

In all three cases, the units are [A], [VA] or [W], and [var].

2.5.5.2 Modelling of the Own Grid Area

Modelling creates abstraction or representations of the system to predict and analyse the performance and provide system research, development, design, and management [46]. Smart Grid modelling involves creating a computer model of the grid that includes all of its components and their properties so that studies can be conducted. Smart Grid elements include generators, lines, transformers, reactive devices, circuit breakers, switches, isolators, household electrical appliances, etc. After all of this information has been entered into the computer, various processes can be carried out, such as monitoring (using SCADA and SE), power flow analyses, Volt/var assessment, etc.

Because Smart Grids are pretty extensive, they are broken down into areas that belong to and are usually operated or used by specific legal entities Fig. 2.36. TSOs, DSOs, and Customers are the actual legal entities involved in Smart Grids. TSOs and DSOs are specialised entities to reliably and economically operate and maintain the grid in their areas. They use SCADA to build the own grid area's static topology, using one- or three-line diagrams. Figures 2.37, 2.38 and 2.39 show the conventional- and *LINK*-modeling TSOs (HV level) and DSOs (MV and LV level) areas.

Figure 2.40 shows the conventional- and *LINK*-modelling of CPs. In the *LINK*-modelling are used conventions discussed above. Although, with the new development, customer plants act as virtual vertical integrated utilities (see Sect. 2.5.1), customers are unprofessional to use the devices following the grid requirements. Therefore, a House Management Unit (HMU) is needed to coordinate relevant facilities with the grid.

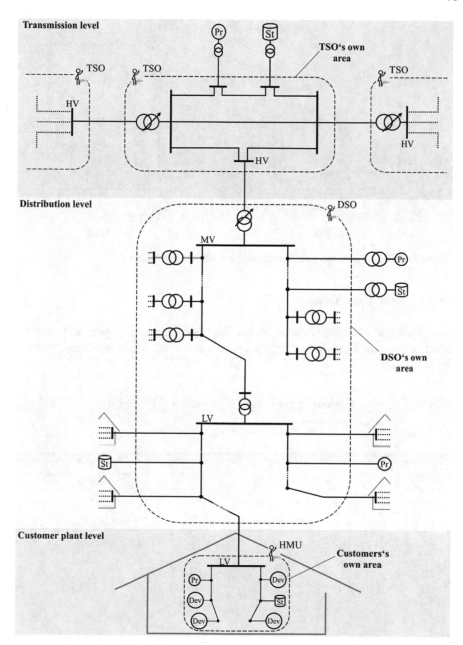

Fig. 2.36 Overview of Smart Grids' parts belonging to and operated or used by specific legal entities such as TSOs, DSOs, and Customers

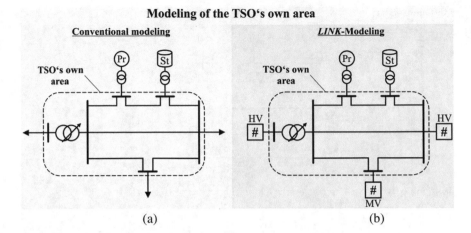

Fig. 2.37 HV level: **a** Conventional modelling; **b** *LINK*-based modelling

2.5.5.3   Grid-Link Types

The Grid-Link definition is quite general. For clarification, some practical cases are listed below based on the traditional grid classification high-, medium-, and low grids.

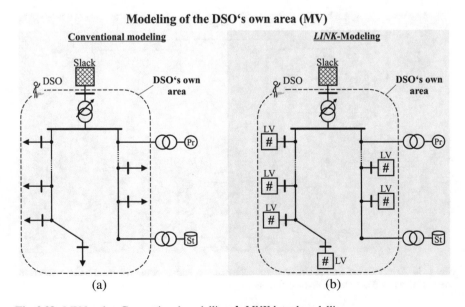

Fig. 2.38 MV level: **a** Conventional modelling; **b** *LINK*-based modelling

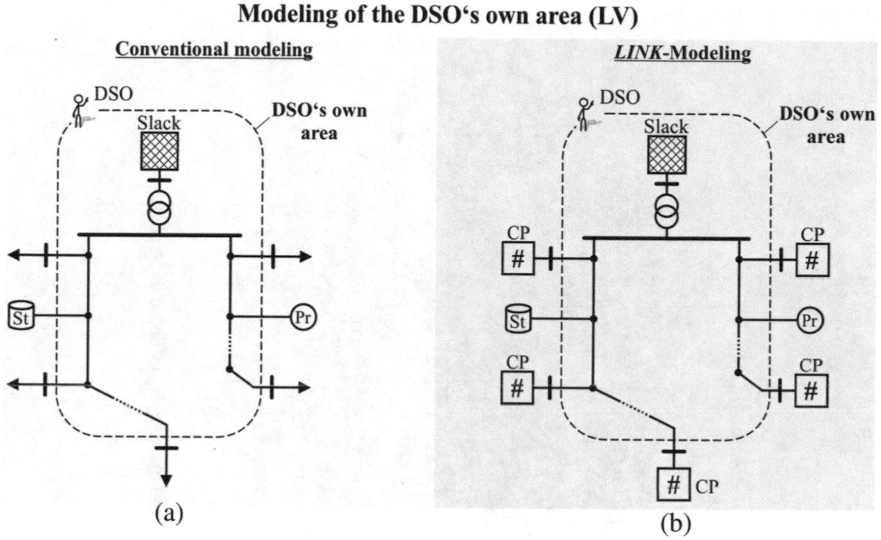

Modeling of the DSO's own area (LV)

Fig. 2.39 LV level: **a** Conventional modelling; **b** *LINK*-based modelling

- **HV_Grid-Link** —The operation/study grid at the HV is usually meshed and contains redundant real-time measurements Fig. 2.41a. The grid's upgrade with Volt/var and Hz/Watt secondary controls and the corresponding interfaces provides the HV_Grid-Link. Figure 2.41b shows the technical/operation architecture of this area, where the Producer- and Storage-Links communicate via interfaces with the HV_Grid-Link. The latter has various HV_ and MV_Grid-Link, external neighbours. That means different system operators operate the latter. The process of day-a-head load-generation balance performed in normal operation conditions realises by exchanging the corresponding schedules, P_{Sched} (i). The SC area may be the same as the TSO control area.

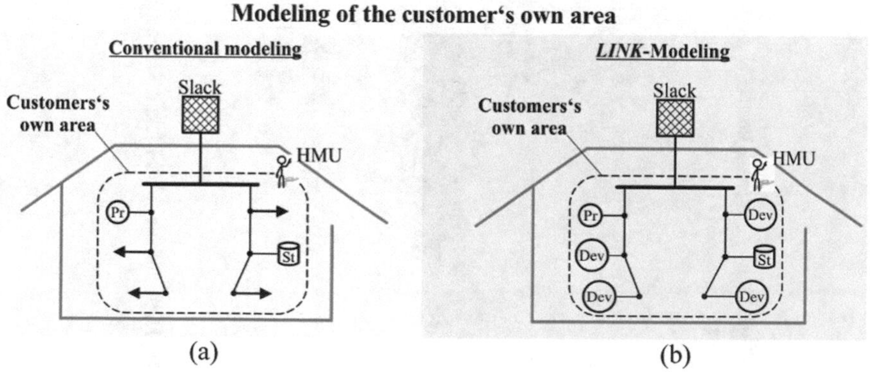

Modeling of the customer's own area

Fig. 2.40 Customer Plant level: **a** Conventional modelling; **b** *LINK*-based modelling

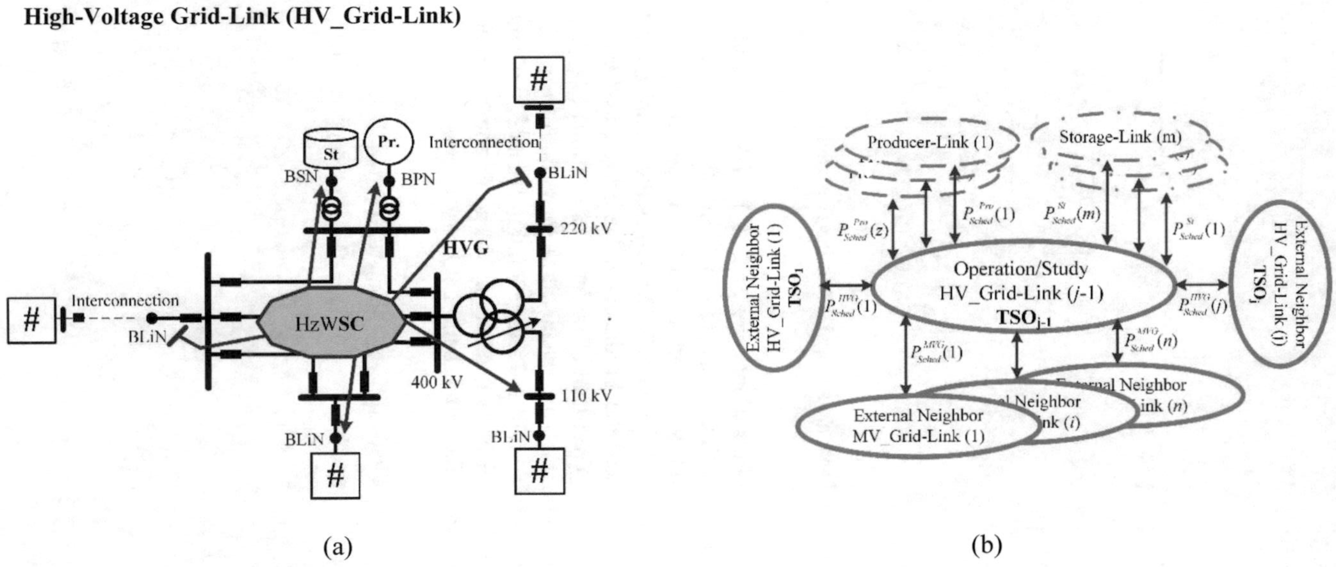

High-Voltage Grid-Link (HV_Grid-Link)

Fig. 2.41 High-Voltage_Grid-Link: **a** Schematic presentation; **b** Functional/operation architecture for the exchanging of day a head schedules

- **MV_ and LV_Grid-Link**—The operation/study grids in MV and LV are usually radial, Figs. 2.42a and 2.43a. The MV grid has one intersection point with the HV grid through the HV/MV supplying transformer, while the LV grid has one intersection point with the MV grid through the MV/LV distribution transformer.

  The upgrades of the grids with Volt/var and Hz/Watt secondary controls and the corresponding interfaces provide the MV_ and LV_Grid-Links. Usually, each SC area will include only one subsystem, which means only one part of the DSO control area. Since the topology is updated almost manually in SCADA, the SC area is dynamically changed, and in the given case, two Grid-Links automatically merge to one. These Grid-Link types differ by the different voltage levels, which impact the electrical parameters of the lines/cables and the availability of the real-time measurements. While the MV_Grid-Links has very few real-time measurements, the LV_Grid-links has none. Figures 2.42b and 2.43b show each area's technical/operation architecture, where the Producer- and Storage-Links communicate via interfaces with the MV_ and LV_Grid-Links, respectively. The latter communicates with the Grid-Link neighbouring via interfaces. The day-a-head load-generation balance process will be performed for each Grid-Link by exchanging the corresponding schedules, $P_{Sched}(i)$.

- CP_Grid-Link sets up over the CP's radial grid. The secondary control's objective function reaches an energetic, economic optimum by fulfilling the agreements/requirements with LV_Grid-Link Fig. 2.44a. In contrast to the other Link types, this Link is not under the administration of utilities but of the customers; HMU takes over controllable devices. Figure 2.44b show each area's technical/operation architecture in the load-generation balance process within the CP by exchanging PSched. CP_Grid-Link communicates via internal interfaces with the Producer- and Storage-Links, and Dev_Grid-Links. It communicates via an external interface with the LV_Grid-Link.

2.5.5.4 Link Interfaces

Each Link or Link-bundle operator is aware of the Producer-Links' electricity-producing capacity feeding into its grid and its limitations. He determines the Link-Grid's topology based on the Producer- and Storage-Links capabilities to reliably supply customers. The information required to enable feasible, reliable, and resilient operation exchanges via interfaces; one of the main elements of every architecture' component. There are three relevant interface pairs the Grid-Link/Grid-Link, Grid-Link/Producer-Link, and Grid-Link/Storage-Link [30]. The relevant electrical entities to be exchanged are shown in Table 2.3.

Grid-Link/Grid-Link Interface

The Grid-Link/Grid-Link Interface is the most extensive one. The day ahead $P_{Schedule}^{dayahead} \pm \Delta P$ and following hour schedule $P_{des}^{nexthour} \pm \Delta P$ exchanges to enable the load-production balance. ΔP is the active power capacity support (spinning reserve),

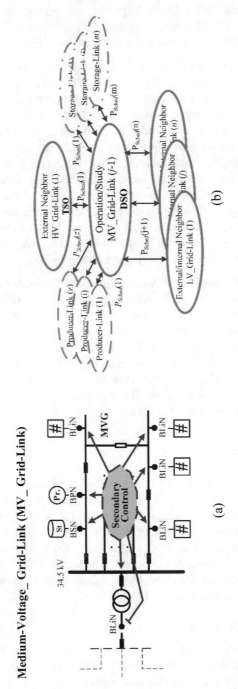

Fig. 2.42 Medium-Voltage_Grid-Link: **a** Schematic presentation; **b** Functional/operation architecture for the exchanging of day a head schedules

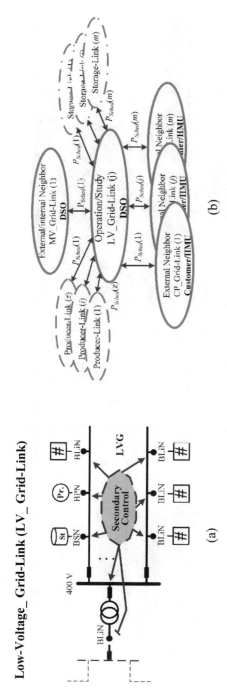

Fig. 2.43 Low -Voltage_Grid-Link: **a** Schematic presentation; **b** Functional/operation architecture for the exchanging of day a head schedules

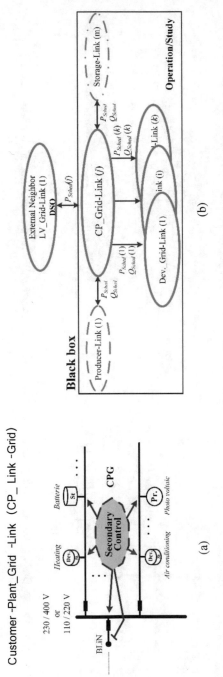

Fig. 2.44 Customer Plant_Grid-Link: **a** Schematic presentation; **b** Functional/operation architecture for the exchanging of day a head schedules

Table 2.3 Electrical entities for all three interface pairs

Electrical entities to be exchanged (*)		Grid-Link/ Grid-Link	Grid-Link/ Prducer-Link(**)	Grid-Link/ Storage-Link
Very fast	f_{meas}	✓	✓	✓
	V_{meas}, δ_{meas}	✓	✓	✓
	P_{meas}, Q_{meas}	✓	✓	✓
	P_{set_point}, Q_{set_point}	✓	✓	✓
Fast	$P_{des} \pm \Delta P$, $Q_{des} \pm \Delta Q$ Delivered time Time interval	✓	✓	✓
	$P_{des}^{nexthour} \pm \Delta P$ $Q_{des}^{nexthour} \pm \Delta Q$	✓	✓	✓
Slow	$P_{Schedule}^{dayahead} \pm \Delta P$ $Q_{Schedule}^{dayahead} \pm \Delta Q$	✓	✓	✓
	Static and dynamic (lumped) load characteristic k_{PV}, k_{QV}, k_{Pf}, k_{Qf} …	✓		
	I_{equiv}, Z_{equiv}	✓		
	Dynamic equivalent Generator parameters like x_d, x'_d, …, T'_{d0}, …	✓	(***)	(***)
	Equivalent voltage regulator, static exciter parameters like K_A, T_A, …	✓	(***)	(***)
	Equivalent governors, turbine parameters like K_1, T_{G1}, …	✓	(***)	(***)
	Schedule for demand response capability	✓		
	Reserves schedule (secondary, tertiary)	✓	✓	✓

*Data related to the boundary node
**P and Q can have only one sign. Producers only inject power on the grid
***Static data should not be exchanged via interfaceHV-HV_Grid-Link

which each Grid-Link should provide during contingency conditions. It is also relevant for the (n-1) security calculations. In this case, also the available reactive power resources should be known $Q_{Schedule}^{dayahead} \pm \Delta Q$ and $Q_{des}^{nexthour} \pm \Delta Q$. The dynamic data for the equivalent generator and the equivalent exciter, voltage regulator, turbine, and the governor are calculated in real-time and exchanged between Grid-Links to enable the angular and voltage stability calculations. Grid-Links can also offer services to each other employing secondary and tertiary reserve. The frequency fmeas is necessary to allow the synchronisation process of Link-Grids, which have been operating in self-sufficient mode. Demand response is the most complicated process of Smart Grids operation. The request for load decrease or increase is included in the interface in the form of the desired instantaneous value, $P_{des} \pm \Delta P$, $Q_{des} \pm \Delta Q$. A detailed description of the Demand Response process is given in Sect. 2.6.5. Depending on the interfering Grid-Link types, the following interface pairs are possible HV-HV_Grid-Link, HV-MV_Grid-Link, MV–LV_Grid-Link, MV- and LV– CP_Grid-Link.

HV-HV_Grid-Link

This Grid-Link interface characterises the typical interaction between, e.g., two TSO areas. Nowadays, one of the crucial HV functions, or rather HV_Grid-Link, is the balance load production in real-time realised via the Automatic Generation Control (AGC) [52]. The HV_Link-Grid practically is the grid part included in the AGC controlling area. AGC controls the real power on a commercial basis. In comparison, the AGC in Link-Grid performs the scheduled interchange obligations to other interconnected utilities (neighbouring HV-Grid-Links) and neighbouring MV_Grid-Links on technical and commercial bases. The frequency and real power, and Voltage and reactive power Secondary Controls, HzW **SC** and VvS**C**, are part of the Grid-Link operating in real-time.

HV-MV_Grid-Link

This Grid-Link interface characterises the interaction between the HVG and MVG taking place through the HV/MV supplying transformer. All entities defined for the Grid-Link/Grid-Link interface should be exchanged to ensure the coexistence of the two Link-types. Both **WSC** and VvS**C** include exchanging over the supplying transformers and providing the setpoints P_{set_point} calculated in real-time. These setpoints have to be treated as dynamic constraints from the MV-Grid-Link' secondary control.

MV-LV_Grid-Link

These Grid-Link interfaces characterise the interaction between the MVG and LVG realised technically through the MV/LV distribution transformer. From the MVG point of view, the LVG was modelled using the lumped feeder load. Similar to above, the MV-Grid-Link' secondary control sends the P_{set_point} and Q_{set_point} to the LV-Grid-Link' secondary control treating them as constraints. Here, the service restoration is more relevant than the $(n - 1)$ security calculation. ΔP and ΔQ are used for restoration purposes. With the increase of the DG share, the analysis of the Grid-Link's dynamic behaviour is relevant.

LV-CP_Grid-Link

This Grid-Link characterises the interaction between the LVG and CPG. The CP_Grid-Link is by definition modular and closed in itself, thus fulfilling the data privacy conditions. Unlike [39], where specific household devices should be turned on/off by network operators and energy suppliers, the CP_Grid-Link acts as a black box in the new functional architecture. The network operator interacts with the CP_Grid-Link through the interface, which gives information only about their exchange and their needs ($P_{des} \pm \Delta P$, $Q_{des} \pm \Delta Q$). No information over the household devices currently in operation is accessible from the grid operator or the energy supplier. The customer may wish to control the house devices by using the Internet of Things (IoT), but this is independent of the customer's interaction with the grid. The communication with the grid takes place only through secure channels, thus protecting the power delivery systems from cyber-attacks. The LV-Grid-Link sends the negotiated set points P_{set_point}, Q_{set_point} to the CP_Grid-Link. At the same time, the CP_Grid-Link's secondary control supervises the real-time exchange with the grid. A powerful HMU generates the daily and hourly P and Q schedules. Theoretically, all entities for the Grid-Link/Grid-Link interface defined in Table 2.3 are necessary also for this link combination. But actually, and for the near future, it is not realistic to collect and prepare this kind of data. Firstly, the house electrical grid is not on the utility nomenclature, and practically they do not have access to it. Secondly, also the customer, as CP_Grid-Link owner, usually does not have the required information. Many research projects show that besides automation, step-voltage regulators should be installed in LV. Over time, this development trend will require more calculations and coordination, and therefore it makes sense to plan this interface with all the data described in Table 2.3.

Grid-Link/Producer-Link interface.

This interface pair characterises the interaction between the TSOs and the power supplier. It is well established in the transmission level, i.e., between the HVG and the electricity producer injecting through step-up transformers. Similarly, the same information exchange between MV_, LV_ and CP_Grid-Links, and the connected Producer-Links.

Grid-Link/Storage-Link interface.

This interface pair characterises the interaction between the TSOs and the storages. It is well established in the transmission level, i.e., between the HVG and the electricity storages (mainly pumped hydroelectric power plants). Similarly, the same information exchange between MV_, LV_ and CP_Grid-Links, and the connected Storage-Links.

2.5.6 Control Chain Net Strategy

LINK-based architecture's standardised structure enables a compact and precise representation of the control strategy, designed as a chain net. The most popular control strategies in power systems are local, primary, and secondary control (see A.2.2). They are mainly used in the transmission part of power systems. The chain net control strategy constitutes a coordinated control set, including direct, primary, and secondary control loops throughout the entire Smart Grid.

> **Secondary control** loops enable the coordination and reliable operation of the Grid-Links in the whole Smart Grid.

Each of the SC loops in the set calculates the corresponding set-points by respecting the dynamic grid constraints. Equation (2.4) presents SC-loops' union, each containing a subset of primary and direct controls and constraints.

$$PpSC_{Chain}^{Axis} = \bigcup_{i=1}^{M} \left\{ PpSC^{Area_i} \left(PC_{Appl}^{Area_i}, DiC_{Appl}^{Area_i}, Cns_{NgbArea}^{Area_i} \right) \right\} \qquad (2.4)$$

where

Pp—the parameter pair Herz/Watt and Volt/var, $Pp \in \{HzW, Vv\}$;
SC_{Chain}^{Axis}—the chain of Secondary Controls in horizontal, X, and vertical, Y, axis. The "Energy Supply Chain Net" holistic model stipulates SC sets in both axes (see Sect. 2.3);
Area—the Grid-Link area where the secondary control is set;
M—the number of secondary controls;
Appl—the appliance such as producer, storage, RPDs, On-Load Tap Changers (OLTC), or Phase-Shifting Transformer (PhST);
$PC_{Appl}^{Area_i}$—Primary controls of all appliances connected at $Area_i$;
$DiC_{Appl}^{Area_i}$—Direct controls of all appliances connected at $Area_i$;
$Cns_{NgbArea}^{Area_i}$—Constraints between the $Area_i$ and neighbouring areas.

Table 2.4 shows the control variables and constraints of chained SC loops. They are classified depending on the relevance for the HzWSC and VvSC loops. Frequency is an area constraint relevant to the HzWSC. The operation active power P of the appliances such as producers and storages connected to the Link-Grids set up the control variables. PQ diagrams of generators and inverters, maximal step number of OLTC for PhST, etc., set up the usual constraints. In the case of VvSC, control variables are the operation reactive power Q of generators, inverters, and RPDs, as well as the switch position of capacitors and coils, the transformer tap step, and so on. Like the HzWSC, PQ diagrams of generators, inverters, maximal step number of OLTCs, installed rating of capacitors and coils, etc., are the VvSC constraints. The electrical

Table 2.4 Relevant control variables and constraints of chained secondary control loops

		Control variables	Constraints
HzWSC	Area		Frequency
	Appliances	Operation value of the active power, P (producers, storages)	PQ diagrams of generators and inverters, maximal step number of OLTC for PhST, etc
	Link-Grid	Active power exchange Pex between neighbouring Grid-Links	Active power exchange P_{ex} between neighboring Grid-Links
VvSC	Appliances	Operation value of the reactive power, Q (producers, inverters, reactive power devices) Switch position of capacitors and coils	PQ diagrams of generators and inverters, maximal step number of transformers, installed rating of reactive power devices, etc
	Link-Grid	Reactive power exchange Q_{ex} between neighbouring Grid-Links	Reactive power exchange Q_{ex} between neighbouring Grid-Links

coordination of neighbouring Grid-Links occurs through the controlled exchange of the active and reactive power P_{ex} and Q_{ex} between them. These exchanging powers make up the dynamic variable controls of the SC chain.

The exchanging active and reactive power P_{ex} and Q_{ex} on the Link-Grid boundaries have a dual nature. They can be converted from grid control variable to grid constraint and vice versa depending on the required action.

2.5.6.1   Secondary Control Chain in *X*-axis

In the *X*-axis, the interconnected very-high and high voltage grids are arranged. Figure 2.45 shows the chain of Secondary Controls in the *X*-axis with details of the HV_Grid-Link *i*. In this Link are stipulated HzW and Vv secondary controls.

HzWSC calculates setpoints for the primary and direct controls of three different types of appliances, i.e., PhSTs, electricity producers and storages, and the SC of two kinds of neighbouring Grid-Links medium and low voltage. The active power flow in tie lines determines the constraints to be respected between the HV_Grid-Links. Equation (2.5) presents the HzWSC chain in the *X*-axis. It is designed as the union of all SCs, including the corresponding subset of primary and direct controls and the SCs of neighbouring Grid-Links.

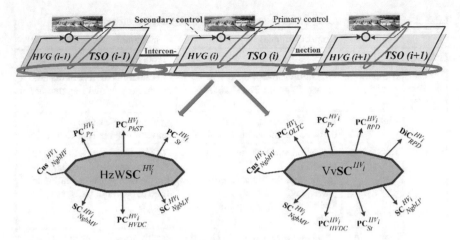

Fig. 2.45 The SC chain in X-Axis with details of the HzWSC and VvSC in HV_Grid-Link i

$$HzWSC_{Chain}^{X-axis}$$
$$=\bigcup_{i=1}^{M}\{HzWSC^{HV_i}(PC_{PhST}^{HV_i}, PC_{Pr}^{HV_i}, PC_{St}^{HV_i}, PC_{HVDC}^{HV_i}, SC_{NgbMV}^{HV_i}, SC_{NgbLV}^{HV_i},$$
$$Cns_{NgbHV}^{HV_i})\}, \tag{2.5}$$

where

M—the number of HV_Grid-Links.

VvSC calculates setpoints for the primary and direct controls of five different types of appliances, i.e., OLTCs, electricity producers and storages, reactive power devices, HVDC facilities, and SCs of two types of neighbouring Grid-Links medium and low voltage. The reactive power flow in tie lines and the voltage in boundary nodes determine the constraints to be respected between the HV_Grid-Links. Equation (2.6) presents the VvSC chain in the X-axis. Similarly to the HzWSC chain, it is designed as the union of all SCs, including the corresponding subset of Primary and Direct Controls appliances and SC of neighbouring Grid-Links.

$$VvSC_{Chain}^{X-axis}$$
$$=\bigcup_{i=1}^{M}\{VvSC^{HV_i}(PC_{OLTC}^{HV_i}, PC_{Pr}^{HV_i}, PC_{St}^{HV_i}, PC_{RPD}^{HV_i}, DiC_{RPD}^{HV_i}, PC_{HVDC}^{HV_i}, SC_{NgbMV}^{HV_i},$$
$$SC_{NgbLV}^{HV_i}, Cns_{NgbHV}^{HV_i})\}, \tag{2.6}$$

2.5.6.2 Secondary Control Chain in *Y*-axis

In the *Y*-Axis, the medium voltage and low voltage and customer plant grids are arranged. They are owned and operated by DSOs and customers, respectively. Figure 2.46 shows the chain of SCs in the *Y*-Axis with details of the Watt Secondary Control (**WSC**) and VvSC in Grid-Link *i*. As the frequency is a global variable, its monitoring and control continue to be carried out by the TSO. The Grid-Links operators on the *Y*-axis, such as DSOs and customers, focus on the Demand-Production balance using **WSC** within their operation area in predefined time intervals. They monitor and control both the voltage and the reactive power using **VvSC** because they are local variables.

    WSC calculates setpoints for two different appliances, such as electricity producers and storages and neighbouring Grid-Links. The coordination with the superordinate Grid-Links realises by the constraints. Equation (2.7) presents the **WSC** chain in the *Y*-axis. It is designed as the union of all SCs, including the corresponding subset of primary and direct controls of appliances and SCs of neighbouring Grid-Links.

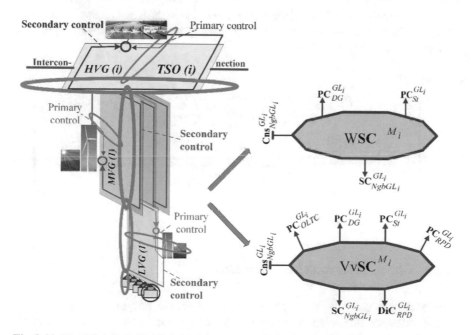

Fig. 2.46 The SC chain in *Y*-Axis with details of the **WSC** and VvSC in Grid-Link *i*

$$WSC_{Chain}^{Y-axis} =$$

$$\bigcup\left\{\begin{matrix} HzWSC^{HV}\left(PC_{PhST}^{HV}, PC_{Pr}^{HV}, PC_{St}^{HV}, SC_{NgbMV}^{HV}, SC_{NgbLV}^{HV}, Cns_{NgbHV}^{HV}\right), \\ \displaystyle\bigcup_{i=1}^{N}\left\{WSC^{GL_i}\left(PC_{Pr}^{GL_i}, PC_{St}^{GL_i}, SC_{NgbGL_i}^{GL_i}, Cns_{NgbGL_i}^{GL_i}\right)\right\} \end{matrix}\right\}$$

$$(2.7)$$

where.
N—the number of Grid-Links in the Y-axis,
GL_i—Grid-Link under examination i; $GL_i \in \{MV, LV, CP\}$,
$HzWSC^{HV}$—the SC set at the HVG in which is connected GL_i,
WSC^{GL_i}—the WSC set at GL_i,
$NgbGL_i$—the Grid-Link neighbour to the GL;
$NgbGL_i \in \{HV, MV, LV, CP\}$.

VvSC calculates setpoints for five different appliances' primary and direct controls, i.e. OLTCs, electricity producers and storages, RPDs, and neighbouring Grid-Links. The reactive power flow between the neighbours and the voltage in boundary nodes determines the constraints to be respected by the VvSCs. Equation (2.8) presents the VvSC chain in the Y-axis. It is designed as the union of all SCs, including the corresponding subset of primary and direct controls of appliances and SCs of neighbouring Grid-Links.

$$VvSC_{Chain}^{Y-axis} =$$

$$\bigcup\left\{\begin{matrix} VvSC^{HV}\left(PC_{OLTC}^{HV}, PC_{Pr}^{HV}, PC_{St}^{HV}, PC_{RPD}^{HV}, DiC_{RPD}^{HV}, PC_{HVDC}^{HV}, \right. \\ \left. SC_{NgbMV}^{HV}, SC_{NgbLV}^{HV}, Cns_{NgbHV}^{HV}\right), \\ \displaystyle\bigcup_{i=1}^{N}\left\{VvSC^{GL_i}\left(PC_{OLTC}^{GL_i}, PC_{Pr}^{GL_i}, PC_{St}^{GL_i}, PC_{RPD}^{GL_i}, DiC_{RPD}^{GL_i}, \right.\right. \\ \left.\left. SC_{NgbGL_i}^{GL_i}, Cns_{NgbGL_i}^{GL_i}\right)\right\} \end{matrix}\right\}$$

$$(2.8)$$

2.5.7 Market Harmonised Structure

A market is a place or an opportunity where people buy and sell goods. The market rules are human-made to enable fair cooperation between market participants. They are subject to constant changes and adoptions depending on the circumstances. Whereas the product "Electricity" differs qualitatively from the usually traded products, its production and distribution must be subject to the electricity's laws, i.e. nature's laws. In contrast to human-made laws, nature's laws cannot be modified—that's why they are so powerful and explosive. For this reason,

> The **market structure** should be designed to conform to the architecture of Smart Grids.

2.5.7.1 *LINK*-Based Market Structure

Although the electricity markets are undergoing a radical change, the current re-dispatch process for congestion management is still costly and drives the transmission grid operation to its limits. On the other hand, the electricity producers connected to the distribution grid cannot participate in the market. The DSOs do not participate in the market operation to manage their congestion management. The effective opera-tion of the Distributed Energy Resources (DER) is quite limited. The transformation of the resource mix of fossil fuels to renewables and the rise of distributed resources call for a radical review of the market structures under a holistic view, including the power grid, customer plants, and the market together.

The holistic market model derived from the technical one (see Sect. 2.4) consists of coupled market areas (balancing groups) at the horizontal and vertical axes and is operated by the TSO and DSO. They interact directly with the whole market to ensure a congestion-free transmission and distribution grid operation. A detailed market structure that fits the *LINK*-holistic architecture is presented below.

The market structure should take into account the technical behaviour of the Smart Grids and support their safe, reliable, and resilient operation while at the same time attracting the demand response bids. Figure 2.47 illustrates the harmonization and coordination of the market structure with the grid link arrangement. The implemen-tation of the local retail markets requires Balancing Group Areas (BGA) within the whole market. The day-ahead market harmonises with the Grid-Links as follows:

- BGA is a geographical area consisting of one or more Grid-Links with standard market rules and has the same price for imbalance in the day-a-head market. In general, a TSO grid includes one grid link, while the grid of a DSO may include several grid links. Additionally, all grid links included in one BGA should be operated by one DSO (not by several DSOs).
  The geographic boundaries of BGA may vary considerably and are defined by the external boundaries of Grid-Links contained in the BGA. Grid links may have two types of boundaries: external and internal. The external ones exist between different links with other owners or operators, i.e., between TSO and DSO. They are subject to data security and privacy because of the data exchanges between two different companies. In contrast, the internal interfaces are set between various links with the same owner or operator, e.g., the same DSO. They are subject only to data security [31].
- All electricity producers, storages, and customer plants electrically connected to the Grid-Links belonging to the BGA participate in the same balancing group.

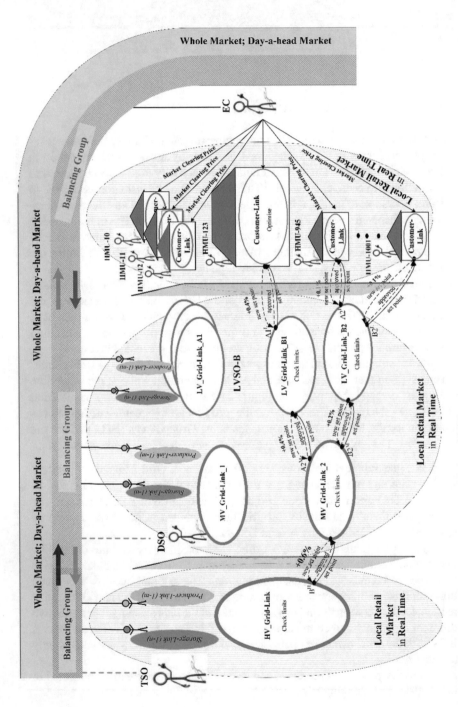

Fig. 2.47 Harmonization of the market structure with the grid link arrangement

> The Grid-Link operators (e.g. a TSO, DSO, or EC) act as neutral market facilitators responsible for controlling and independently balancing power flow fluctuations in their area.

Moreover, each Grid-Link operator provides ancillary services to the neighbouring Grid-Link areas to ensure the reliability of the electricity supply.

2.5.7.2 *LINK*-Structure of the Austrian Electricity Market

LINK-Solution stipulates the communication of Links with the electricity market through the market interface "M" (see Sect. 2.5.3.4). The illustration of market interface "M" is based on the Austrian electricity market [12]. Some of the most prominent participants in this market are:

- Control Area Manager (CAM) → responsible for load–frequency control within its control area.
- System Operator (SO) → operator of transmission or distribution grid.
- Supplier → commercial provider of electric energy
- Consumer → buyer of electric energy for own use.
- Clearing and Settlement Agent (CSA), i.e. balance group coordinator → responsible for organising, clearing, and settling of balance within a control area.
- Balance Responsible Party (BRP), i.e. balance group representative → represents a balancing group vis-à-vis other market participants and vis-à-vis the CSA.

In the Austrian electricity market, balance groups are introduced to enable consumers, generators, suppliers, and wholesalers to trade or conclude deals with each other. Whoever takes electricity off the power grid, feeds in, or trades must be a member of the balance group. BRP balances the load within the balance group. CSA takes care of load balancing within the control area, including more than one balancing group, up to the intraday trade. The CAM ensures the system stability in real-time by performing the load–frequency process, which in reality is handled by TSOs.

Consequently, TSO assumes a new role, the one of CAM. The number of the data that should be sent to CAM or rather to TSO is immense ([32]; [12]). Besides the different contract types—i.e., utilisation-, supply- or storage contract -, meter readings, and bills, market participants exchange schedules for electricity trading. Here it should be noted that there are two types of schedules:

- **Internal**: Internal schedules for electricity trading between balance groups in the same control area.
- **External**: External schedules for electricity trading between balance groups in different control areas.

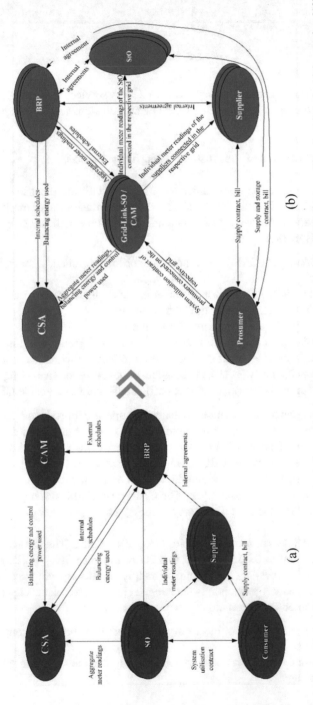

Fig. 2.48 Schematic presentation of contractual relations and information exchange among market participants: **a** The Austrian electricity market [12]; **b** *LINK*-structure of the Austrian electricity market

Figure 2.48 shows a schematic presentation of contractual relations and information exchange among market participants: (a) of the Austrian electricity market model; and (b) its transformation conforms to the *LINK*-Solution. It depicts the contractual relations and information exchange among market participants. Market participants and their roles remain almost the same as in [12] up to the TSO and DSO's roles combined with CAM. *LINK*-Solution introduces three crucial novelties in the electricity market model:

1. Storage Operator

Storage operator emergence is a logical consequence of the new technological developments (see Sect. 2.5.4.1).

2. Consumers transform into prosumers

In recent years, many consumers have gradually been transformed into prosumers. Their inclusion in the market model is essential for the equitable development of the electricity market.

3. Adoption of the CAM role.

Adopting the CAM role is crucial for a secure, reliable, and efficient power system operation by fulfilling data privacy and cybersecurity requirements. In the new market model, each of the Grid-Link-SO, i.e. TSO and DSO, assumes the new role CAM. That means that each Grid-Link-SO also has the CAM role and is responsible for the load balance process in their control area up to the real-time time frame. The technical realisation is realistic because a secondary control for the active power and frequency is designed per each Grid-Link type.

In this new electricity market structure, the scheduling process is stipulated to remain the same as described in [12]. The only difference is the exchange of the pumping with storage schedules.

The new market structure harmonises with the Grid-Link arrangement, alleviates the typical re-dispatch process, and facilitates the most challenging operation process of Smart Grids, the demand response process.

2.6 Chains of Smart Grid Operation

The main goals of Smart Grids operation are safety, reliability, and efficiency. The Smart Grids operation is particularly complex and challenging because the renewable energy sources dominate through the whole power system and customer plants. It affects human safety, supplying reliability, ecological footprint, and operating costs associated with electricity production and storage resources.

As [69] pointed out, the Grid-Link operation also has three cornerstones:

- **Operator**, who is responsible for executing different processes to guarantee a safe, reliable, and efficient operation;
- **Process**, which provides detailed execution steps under various conditions. The most fundamental processes are:

- – Monitoring,
- – Production load balance,
- – Volt/var control,
- – Congestion alleviation,
- – Reserve monitoring,
- – Static and dynamic stability,
- – Power recovery, and,
- – Demand response etc.

- • **Technology** or decision support tools, which enables and facilitates the processes.

In other words, the Grid-Link operation involves creating a picture of the prevailing Grid-Link(s) operating conditions by measuring various system quantities such as frequency, voltages, flows, etc. Based on this picture, the Grid-Link operator initiates and executes different processes based on the predefined steps (e.g., described in the operating order book) supported by the related technology. This procedure ultimately leads to automated control actions by the management systems or manual-remote control actions by the Grid-Link(s) operator or manual by the field crew, as shown in Fig. 2.49.

To ensure a satisfactory operation of Smart Grids, *LINK*-Solution postulates a chain operation of Grid-Links (see Sect. 2.6). The different operation chain processes are examined in the following.

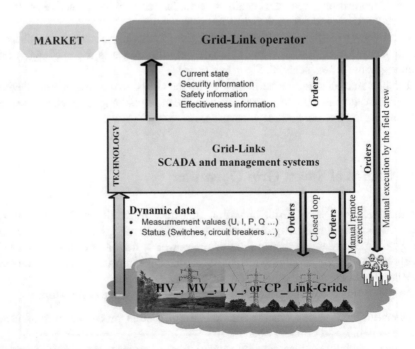

Fig. 2.49 Grid-Link operation process

2.6.1 *Monitoring*

Link-Grid monitoring is the central process of Smart Grids operation. It includes the investigation of Link-Grid conditions. Operators examine the prevailing system conditions to determine whether the grid operates within the physical and operational limits defined primarily in Grid Codes. Frequency, voltages, currents, active and reactive powers, circuit breakers and switches position status, etc., are some monitoring quantities the operator observes to evaluate the operating situation.

The monitoring quantities are usually measured in real-time, which requires the implementation of costly technologies. Figure 2.50 shows the structure of the traditional power system with real-time measurements. Since HVGs are the backbone of the power grid, TSOs make significant investments in technology to ensure a safe and reliable power supply. Therefore, HVG is equipped with redundant measurements and remotely controlled circuit breakers, and sophisticated management systems.

In contrast, MVGs and LVGs were the less glamorous part of power grids. The electricity flows in one direction, and their operation was mainly manageable without using any noticeable technology. The rise of distributed generation and renewable energy sources completely changes the situation. The electricity flows bi-directional and gains a high dynamic due to the volatile behaviour of the renewable energies.

LINK **Solution** postulates monitoring in every Grid-Link.

SCADA is the technology used for measuring, transporting, and finally displaying the information. The SE application supports the operator with error- and gap-free information.

2.6.1.1 SCADA

SCADA is a technology that enables the user to monitor multiple remote facilities by gathering data and even supervise the control operations. It relies on computers, software, and communication media [5].

SCADA systems are widely used in power systems. Figure 2.51 schematically presents a SCADA system in an HV_Link-Grid. It comprises three parts:

- Enabling of data acquisition and orders execution, i.e. change tap position, change circuit breaker position;
- Communication infrastructure; and.
- Supervisory control.

The Remote Terminal Unit (RTU) interfaces the objects in the physical world to SCADA by transmitting telemetry data to a master system and using messages from the master supervisory system to control connected objects [7]. It interfaces the central control unit using standard protocols in various communication media, e.g.

Power system structure:

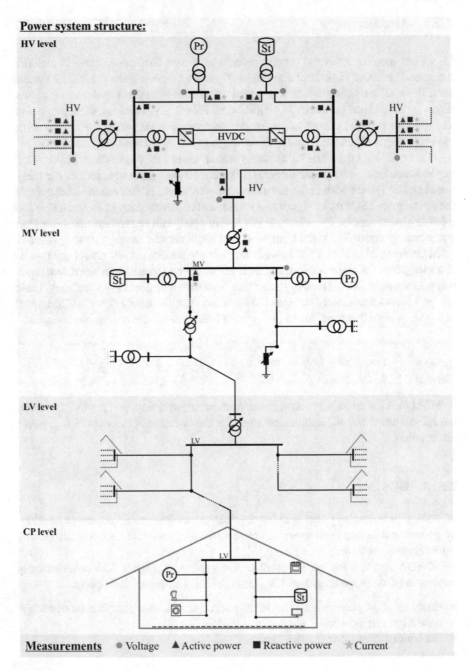

Fig. 2.50 Structure of the traditional power system with real-time measurements

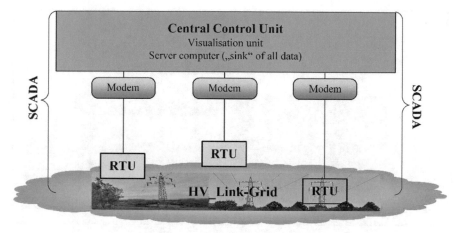

Fig. 2.51 Schematically presentation of SCADA system in an HV_Link-Grid

ethernet. RTUs are usually placed in substations and collect the appliances status and the monitoring quantities such as voltages, active and reactive powers, currents, etc. The central control unit continuously collects the information gathered by RTUs in its database and visualises it.

2.6.1.2 Steady-State Estimator

The steady-state of an electric power system is defined as the vector of the voltage magnitudes and angles at all grid buses. The static-state estimator is a data-processing algorithm for converting redundant meter readings and other available information into an estimate of the static-state vector [63]. After about 20 years, a state estimation algorithm for medium voltage grids with minimal quantities on real-time measurements was developed and further improved [10, 57].

Figure 2.52 shows the state estimator and relevant applications embedded in the SCADA system in an HV_Grid-Link. Using an incomplete and inconsistent set of measured values, the state estimator calculates a complete and consistent set of estimated values, creating the foundation for other applications such as Volt/var control, optimal power flow, etc.

2.6.2 Volt/var

Reactive power is a by-product of an Alternating Current (AC) system generated and absorbed by various power grid appliances. It is caused by overexcited synchronous machines, capacitors, cables, lightly loaded overhead lines, inverters, and sank by under-excited synchronous machines, induction motors, inductors, transformers,

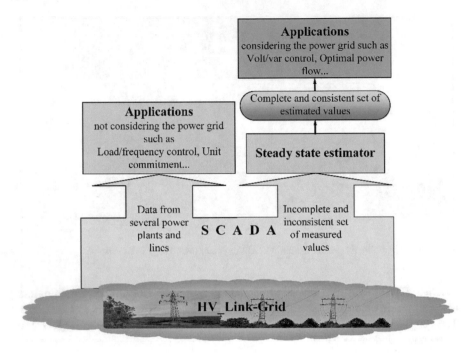

Fig. 2.52 State estimator and applications embedded in the SCADA system in an HV_Grid-Link

heavily loaded overhead lines, and inverters. Transformers and rotating machines need reactive power to create the magnetic flux, and at the same time, it is an essential lever to control the voltage in long cables and overhead lines. That's why Voltage, Volt, and the reactive power, var, are always considered together. Their optimised exploitation achieves a more efficient gird operation by reducing system losses, peak demand, energy consumption or combining the three.

Volt/var control is an essential process in the operation of a power grid and, consequently, Smart Grids. In *LINK*-Solution, the Volt/var control or management take place in each Grid-Link. The control loops are outlined on each Grid-Link, creating a Volt/var interaction chain [31]. Figure 2.53 shows a schematic presentation of the resilient VvSC chain in the Y-axis of a typical European power system. The transmission grids are under the administration and operation of TSOs, while the sub-transmission and the distribution ones under DSOs. Each of them has the own operation control centre. An HVT_, HVS_, MV_, LV_, and CP_Grid-Link is set on the transmission (very high and high voltage grid), sub-transmission, medium, low voltage grids, and in customer plant grids, respectively, Fig. 2.53a.

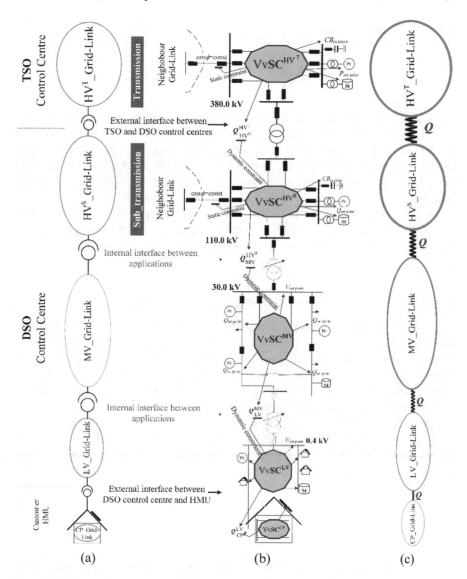

Fig. 2.53 Schematic presentation of the resilient VvSC chain: **a** Link structure; **b** Volt/var control loops; **c** Resilient connection through the reactive power

2.6.2.1 Resilient Interaction of VvSC Loops

In terms of network management, two types of interfaces are identified: external and internal. The external interfaces locate between different Links, having other owners, i.e., between TSO and DSO. They are subject to data security and privacy because of the data exchanges between two different companies, Table 2.5. Internal interfaces

Interface type	Data security	Data privacy
External	$\checkmark$	$\checkmark$
Internal	$\checkmark$	

Table 2.5 Different interface types in *LINK*-Solution

are set between different Links, having the same owner. In this case, the DSO. They are subject only to data security. That means interfaces between HV^T_Grid-Link and HV^S_Grid-Link are external, while interfaces between the HV-, MV_ and LV_Grid-Links are internal. The interface between LV_ and CP_Grid-Link is external, as the information exchange occurs between the DSO and the customer, specifically HMUs.

Figure   2.53b   shows   the   VvSC   loops   of   each   Link: $VvSC^{HV^T}$, $VvSC^{HV^S}$, $VvSC^{MV}$, $VvSC^{LV}$, and $VvSC^{CP}$. Each VvSC application calculates the relevant set points by optimising its own decisions that are subject to:

- Its constraints; and,
- Dynamic constraints imposed by neighbouring Grid-Links.

Dynamic constraints control the reactive power flow in a VvSC chain [34]. They are called dynamic constraints because they change or should be recalculated in real-time depending on the current situation. For example: if the reactive power $Q_{MV}^{HV^S}$ supplied from HV^S_Grid-Link into the MV_Grid-Link should be reduced by 40%, a new desired constraint is sent to the MV_Grid-Link. VvSC^{MV} recalculates the set points in its area by respecting the new condition with the superordinate grid.

Otherwise, if the actual $Q_{MV}^{HV^S}$ is not optimal for the MV_Grid-Link operation, a request is sent to the HV^S_Grid-Link to change it, and so on. The same schema works across the entire VvSC chain. This permanent exchange of desired reactive power Q between different VvSC loops creates a resilient interaction between them, Fig. 2.53c.

2.6.2.2   Working Principles in a Chain

The dynamic grid constraints are introduced to enable a resilient interaction of VvSC loops on the chain. They have a dual nature as they can be converted from grid constraints to grid control variables depending on the action to be performed.

This dual nature is illustrated through the use-case voltage monitoring and control, Fig. 2.54. HV^T_Grid-Link is connected to BLiN-B of HV^S_Grid-Link via the BLiN-A. Between them flows $Q_{HV^S}^{HV^T}$. Figure 2.54a shows the system operation in the presence of a contingency; Voltage violation in the sub-transmission grid. The $VvSC^{HV^S}$ application runs to find an adequate solution without violations: The actual $Q_{HV^S}^{HV^T}$ constraint is not optimal for the HV^S_Grid-Link operation. The application relaxes the operation constraint on point B by changing the status of $Q_{HV^S}^{HV^T}$ from grid constraint to the grid control variable. The $Q_{HV^S}^{HV^T}$_new is calculated, and a change

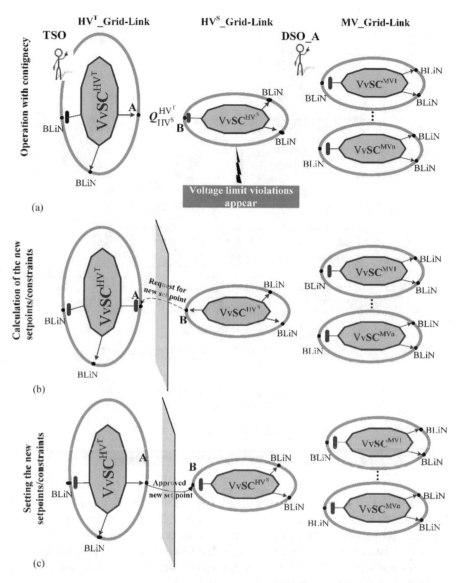

Fig. 2.54 Use case-voltage violations in the sub-transmission grid: **a** Operation with contingency; **b** Calculation and change request of the new set-points/constraints; **c** Approving and setting of the new setpoints/constraints

request is sent to HV^T_Grid-Link, Fig. 2.54b. HV^T_Grid-Link checks the new request by treating the $Q_{HV^S}^{HV^T}\_new$ like a constraint and vice versa; The grid controls are set in dynamic lists. Figure 2.54c shows the approving and setting process of the new desired set point. The same procedure is used to alleviate the violations or optimise the operation in other Link types (other grid parts).

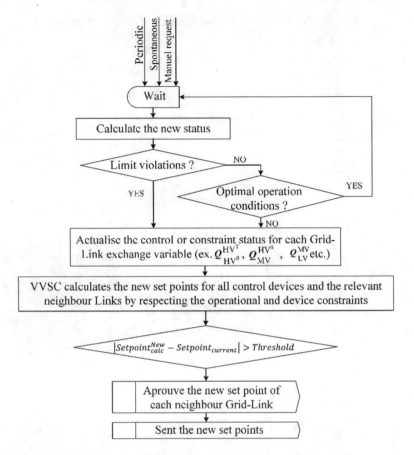

Fig. 2.55 Algorithm of the generalised, concatenated Volt/var secondary control

LINK-Solution enables the generalization VvSC algorithm. Figure 2.55 shows the algorithm of the generalised, concatenated VvSCs. One of the three triggers, periodic, spontaneous, or manual-request, initiates the calculation of the grid state that in the best case is a state estimator [37]. The latter verifies the operating limits. If limit violations are identified, the VvSC application starts calculations after updating the dynamic lists of grid controls.

2.6.3   Hz/Watt

Historically, TSO was the only operator to conduct the load-generation balancing process. The load represents the summation of native loads and losses of the own area and the scheduled exchanges to other areas (other utilities). The load-generation process occurs in three different time frames; sub-processes [69], Fig. 2.56.

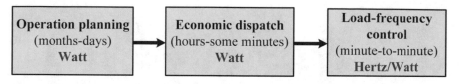

Fig. 2.56 Load-generation balance process broken down into three sub-processes

Sub-process operation planning deals with changes in the transmission grid or generation that need to be made for maintenance purposes in the coming months. The time frame is large because generation units need different preparation times to be ready for the operation, ranging from months to days. For example, while nuclear plants need months of preparation, hydraulic plants can be commissioned within a day or so. Sub-process economic dispatch is about selecting the most economical units to supply the load in the next few hours. Meanwhile, the sub-process Load Frequency Control (LFC) is to balance generation and load on a minute-to-minute basis.

> In the **_LINK_-Solution**, the load-generation balance process takes place in every Grid-Link or Grid-Link bundle. Every Grid-Link operator (TSO, DSO, or customer) conducts the load-generation balancing process in its Grid-Link or Grid-Link bundle. The load represents the summation of the scheduled exchanges to neighbouring external Grid-Links.

Figure 2.57 shows a schematic presentation of the Grid-Links in the Y-axis with the corresponding HzWSC chain. Since frequency is a global parameter of power systems, the sub-process load–frequency control that happens minute-to-minute is attributed only to HV^T_Grid-Links (set up on today's transmission grids). $HzWSC^{HV^T}$ is stipulated for this Grid-Link. Meanwhile, both sub-processes, the operation planning and economic dispatch apply in all Grid-Links. Consequently, the load-generation balance process in Grid-Links in distribution and customer plant levels includes only the operation planning and economic dispatch sub-processes. For them, only **WSC** is stipulated. Figure 2.57b shows the HzWSC loops of each Link: $HzWSC^{HV^T}$, WSC^{HV^S}, WSC^{MV}, WSC^{LV}, and WSC^{CP}. Similar to VvSC, each SC application calculates the relevant set points by optimising its own decisions that are subject to:

- Its constraints; and,
- Dynamic constraints imposed by neighboring Grid-Links.

Dynamic constraints control the active power flow in a HzWSC chain. They change or should be recalculated in real-time, depending on the current situation. For example: if the active power $P_{MV}^{HV^S}$ supplied from HV^S_Grid-Link into the MV_Grid-Link should be reduced by 20%, a new desired constraint is sent to the

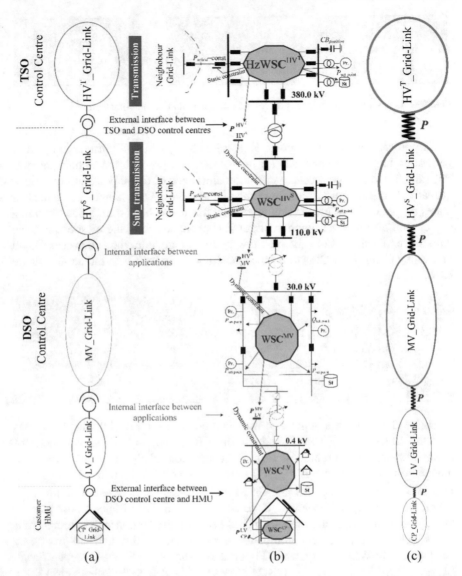

Fig. 2.57 Schematic presentation of the resilient HzWSC chain: **a** Link structure; **b** Hertz/Watt control loops; **c** Resilient connection through the active power

MV_Grid-Link. WSC^{MV} recalculates the set points in its area by respecting the new condition with the superordinated grid. Otherwise, if the actual $P_{MV}^{HV^S}$ is not optimal for the MV_Grid-Link operation, a request is sent to the HV^S_Grid-Link to change it, and so on. The same schema works across the entire HzWSC chain. This permanent exchange of desired active power P between different HzWSC loops creates a resilient interaction between them, Fig. 2.57c. The working principles of the HzWSC chain are similar to the VvSC chain (see Sect. 2.6.2.2).

2.6.4 Reserve Monitoring

Traditionally, the reserve monitoring process took place in transmission utilities. It observed the active and reactive power reserve capacities needed within the utility, taking into account the interchanges with other utilities. The capacity reserve sustains the electricity system's reliability by ensuring more supply than demand is available. Suppose the system has the generation capacity equal to the electricity demand. In that case, an electricity shortage may occur when just one power plant cannot operate as usual, or there is a sudden demand increase.

In the *LINK*-Solution, the reserve monitoring process takes place in every Grid-Link or Grid-Link bundle.

> Every **Grid-Link operator** (TSO, DSO, or customer) conducts the reserve monitoring process in its Grid-Link or Grid-Link bundle. They observe the reserve capacities on active and reactive power needed within the utility, taking into account the reserved capabilities declared by neighbouring external Grid-Links.

LINK-Solution enables the demand response process that, additionally to the supply side, allows the active participation of the demand side to allocate reserve capacities.

2.6.5 LINK-Flexibility: Demand Response

Flexibility and robustness are core design principles of the *LINK*-Solution.

> **Flexibility** means having the possibility to overcome every challenge appearing during the operation of Smart Grids by maintaining the quality of parameters of the supply. It ensures a robust infrastructure, providing its services even during exceptional events.

Smart Grids with the holistic *LINK*-architecture support the flexible operation of electricity producers, storages and customer plants by enabling the demand response.

Demand Side Management (DSM) and DR are processes that try to modify the electricity consumption shape of customers. DSM was coined following the 1970s energy crisis, and since then, it is continuously used by electricity utilities as an instrument for increasing efficiency and shaving peaks [20]. It includes almost medium to long-term countermeasures. With the technology progress and the rise of distributed generation are opened other perspectives, on-demand shifting and the reduction of total energy consumption. DR rose and dedicates to short-term load reduction in response to a signal from the power grid operator or a price signal from the electricity market. Nowadays, DR is being introduced very slowly, especially in the residential, commercial, and small business sectors [64]. The proposed structures are pretty complicated and require significant data exchange [16, 56] that provokes a remarkable increase in the complexity of system operation.

The *LINK*-architecture allows the proper launch of the emergency and price-driven DR. In the case of price-driven DR, the demand change is triggered by a price signal from the electricity market. While, in the case of emergency DR, the Grid-Link operator triggers, e.g. the demand reduction to alleviate overloading in the own area.

The activation of residential, commercial, and small business sectors, which join the real-time pricing demand response process through already concluded contracts, may be triggered at any time. Their degree of participation in the demand response process may be different depending on the time of the day, duration interval, price value, etc.

2.6.5.1 Price-Driven

The electricity price in the market decreases by electricity surplus. All market participants and market operators are notified so that they can act on time. Figure 2.58 shows the information flow during price-driven demand response. It enables residential, commercial, and small business sectors to perceive transparent energy prices and contribute to the reliable and efficient operation of Smart Grids. Let's assume that conditioned from weather and the minimal load consumption, the electricity surplus in the market provokes a price decrease. Figure 2.59 shows the information flow in the CP during price-driven demand response. The new, reduced price is sent via the aggregator or EC through the market interface "M" to the HMU of each customer [33]. After checking the possibilities of demand increase in CP, HMU-123 requests to increase the consumption by 0.4% via the technical interface T to the BLiN $A1^L$ of LV_Grid-Link_1.

Consequently, the Low Voltage System Grid-Link Operator-B (LVSO-B) checks power flow limits under the new conditions at the own Grid-Link. If the power exchange in the BLiN $A2^M$ with the MV_Grid-Link_2 is affected, LVSO-B should pass over the request to the Medium Voltage System Link-Grid Operator-A (MVSO-A). After collecting all incoming requests, MVSO-A calculates power flow in the own Grid-Link. He requests the High Voltage Grid-Link System Operator (HVSO) for a flow increase of 0.6% in the BLiN B^H. HVSO collects all incoming requests and performs the necessary calculations, i.e. power flow, n-1 security, etc. If no limits are violated, HVSO approves the new setpoints and notifies the MVSO-A. The last one accepts the new set points in BLiNs $A2^M$ and $B2^M$ and notifies LVSO-B. He approves

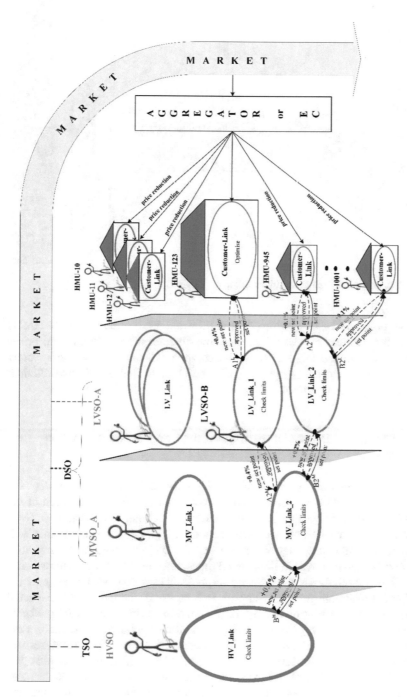

Fig. 2.58 Information flow of the price driven DR process

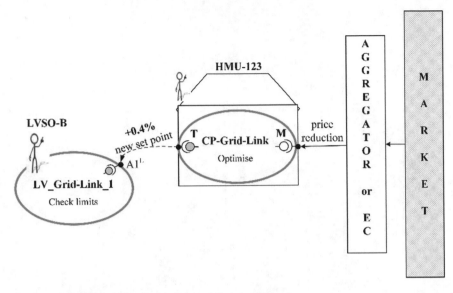

Fig. 2.59 Information flow of price-driven DR at the CP level

the new set points in BLiNs A1^L, A2^L, and B2^L, and notifies the respective HMUs to execute the demand increase. The one flow diagram of emergency- and price-driven demand response enables residential, commercial, and small business sectors to perceive transparent energy prices and contributes to the reliable and efficient electric power system operation.

2.6.5.2 Emergency Driven DR to Support the Congenstion Alleviation Process

The current structure and operation of the electricity market result in a re-dispatch process for congestion management that is costly and pushes the operation of the transmission system to its limits. In this process, one thing that matters to TSOs is whether active power flows decrease or increase at the intersection points with DSOs without being in how they occur. Whereas the DSOs are required to keep voltages throughout the grid within limits at all times; also during the demand response process. Figure 2.60 shows the demand response [1] process used to support the congestion management process. Suppose an increase in the overload is expected in a high-voltage transmission line up to 8% in the following hours. TSO performs the congestion alleviation process: Using the relevant applications, he defines the BLiNs A^H and B^H on his grid where the load decrease should be 2% and 6%, respectively. Both Grid-Links connected on the BLiNs are MV_Grid-Links. They are operated from the same operator DSO_A. Afterwards, TSO initiates a demand decrease request and proposes two new setpoints accompanied by the setting and duration.

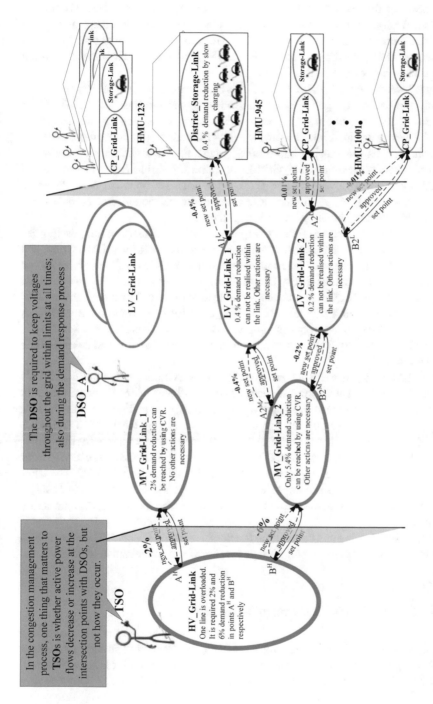

Fig. 2.60 Information flow of emergency-driven DR process used to support the congestion alleviation process

After receiving the request for the new setpoints, DSO_A starts the demand response process and investigates all possibilities to realise the demand decrease using their internal resources, e.g., the CVR [40, 43]. The 2% power reduction in the BLiN A^H was realised by performing the CVR on MV_Grid-Link_1. No other actions are needed. The new setpoint is notified to TSO.

The reduction desired on the BLiN B^H is more extensive than at A^H, about 6%, and only one part of it, e.g. 5.4%, can be reached by performing CVR in MV_Grid-Link_2. For the rest, about 0.6% demand reduction, other actions are necessary. DSO_A investigates his Link-Grid and the day-1 schedules and identifies the BLiNs $A2^M$ and $B2^M$ as the most suitable ones, where the flow should be reduced by 0.4% and 0.2%, respectively. LV_Grid-Link_1 and LV_Grid-Link_2 are connected respectively to the BLiNs $A2^M$ and $B2^M$. Both links are operated from the same DSO. Afterwards, DSO_A initiates a demand decrease request and proposes two new setpoints accompanied by the setting and duration.

After receiving the request for new setpoints, DSO_A investigates all possibilities to realise the demand decrease. He cannot perform the CVR in its Link-Grids, and therefore, he should pass over the request to the customers, who already have signed a contract for participation in the DR process. After performing the calculations, DSO_A finds three BLiNs that are most suitable to realise the demand reduction: $A1^L$ in LV_Grid-Link_1, and $A2^L$ and $B2^L$ in LV_Grid-Link_2. Consequently, DSO-A initiates a demand decrease request and gives over the load decrease of 0.4%, 0.01%, and 0.01%, respectively. The request is accompanied by the setting and duration time of the new setpoints.

HMU-123, which is a district garage with e-cars, is connected to the BLiN $A1^L$. After receiving the new setpoint request, HMU-123 investigates all possibilities to realise the demand decrease. He approves the new setpoint and notifies the DSO-A. The same approval and notifying procedure is also used by HMU-945 and HMU-1001. After collecting all replies, DSO-A agrees on the new setpoints for the BLiNs $A2^M$ and $B2^M$. Having the approvals from all relevant BLiNs, DSO_A can also fulfil the BLiN B^H requirements, approve the new setpoint, and inform the TSO. The latter sent the ultimate set points accompanied by the setting and the duration time. DSO_A makes the final changes on the setpoint schedules and sent the information further up to HMUs.

Thus, by supervising and controlling the fluxes at the BLiNs, the SC chain enables the cross-demand response by all voltage level grids up to the native load.

2.6.6 LINK-Flexibility: Power Recovering

With the associated chain of secondary controls, the Grid-Link structure enables a flexible distributed operation of the Smart Grid for recovery after a large blackout. Figure 2.61 describes the recovery process after a large blackout on a sunny day.

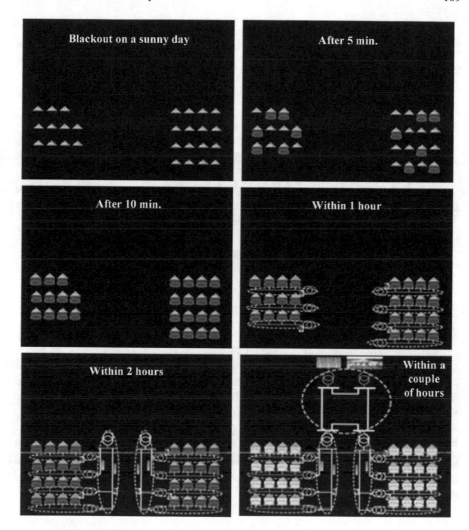

Fig. 2.61   Recovery steps after a blackout

<u>Blackout on a sunny day</u>
  All customer plants with installed PVs have the opportunity to supply individually at least their minimal load by changing the operation mode from normal-autonomous to autarkic-recovery.

After 5 min

Within 5 min, some of the customer plants with PV have set the operation mode to autarkic-recovery and have partially supplied the load.

After 10 min

Almost all customer plants with PV have set the operation mode autarkic recovery and have partially supplied the load.

Within one hour

In each LV_Grid-Link is set the operation mode to autarkic-recovery, thus supplying partially the customer plants without PV installation.

Within two hours

In each MV_Grid-Link is set the operation mode to autarkic-recovery, thus supplying partially all customer plants, which could not be supplied individually or by the LV_Grid-Links. At least the minimal load of all customers is supplied.

Within a couple of hours

The HV_Grid-Links have recovered, and the full load is supplied.

2.6.7 Static and Dynamic Stability

Figure 2.62 shows the dynamic security process for the HV_Grid-Link when a new DG is switched-on on the grid of the MV_Grid-Link_2. Although the DG is not part of HV_Grid-Link, it impacts its dynamic behaviour. Therefore, the new parameters [44] for the Dynamic Equivalent Generator (DEG) DEG^{new} and the equivalent impedance EI^{new} related to the B^M will be calculated online. The new calculated values will be committed to the HV_Grid-Link (BLiN, B^H) if they are different from the old ones Fig. 2.62a. Thus the HV_Grid-Link is notified that one of the neighbours has changed the dynamic behaviour. For this reason, HV_Grid-Link will initiate the calculation of the dynamic stability (angular and voltage) of its Link-Grid with the updated parameters on the calculation model, Fig. 2.62b.

2.7 Protection

Sometimes power systems are subject to abnormal conditions due to device failures, natural events, or human errors in operation. Protection relays and related systems are devices, usually installed on circuit breakers, used throughout the power system to detect and isolate these abnormal conditions. There are many strict requirements to avoid the latter because they can have quite significant consequences. Therefore, protection systems in the context of the *LINK*-Solution must be upgraded so that they can continue to perform their functions with a very high degree of reliability and security.

For example, the protection relay design for the LV_ and MV_Grid-Links should be different from the one used in the past for medium and low voltage grids. Links

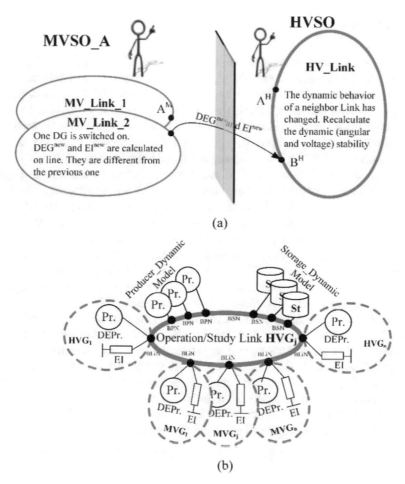

Fig. 2.62 Dynamic security process for the HV_Grid-Link: **a** Interlink information exchange; **b** Calculation model

allow the highest DER penetration and are subject to the bidirectional flows and two modes of operation: autonomous and autarkic. The bidirectional flows become the norm in most radial configurations. In addition, a significant change in short circuit capability occurs when Grid-Links switch from autonomous to the autarkic mode of operation. The last operation mode will profoundly impact the vast majority of protection schemas based on short-circuit current detection.

In summary, the *LINK*-Protection should cover both operating modes, autonomous and autarkic. Setting it up requires a thorough study, which will be presented in a separate chapter of a subsequent edition of this book.

2.8 Planning

The holistic architecture requires the close integration of operation and planning. Hence, future grid design standards, supported by an appropriate regulatory framework, should recognise the added value of the *LINK*-based holistic architecture. For example, the definition of capacity in the future standards grid design should consider the capability of emergency loading of network assets because of the facilitation of the demand response process. Thus, they gain the ability to provide additional capacity in the short term and reduce the amount of demand to be interrupted. It may be cost-effective to increase the life-loss of the assets by overloading these during emergency conditions, as most of the time, the assets are operated below the nominal rating.

The key issue regarding the future evolution of the electricity network design standards is associated with the efficiency of the operational strategies. The latter is critical to determining how much network capacity should be released to grid users under various conditions and how advanced off-grid assets and technologies could support this capacity release. There is a clear trend in using advanced technologies that can provide sufficient security through a more flexible and sophisticated system operation, rather than through asset redundancy only. In this context, the implementation of the *LINK*-Architecture supports the reduction of network redundancy. It ensures the security of supply by enabling the application of a range of advanced, technically effective, and economically efficient corrective actions (or post-disruption actions) that can free up the latent network infrastructure capacity of the existing system. v-Solution enables higher utilisation of the current network assets without compromising the reliability of supply.

Planning with the *LINK*-Architecture in mind requires a thorough investigation, presented in a separate chapter of a later edition of this book.

2.9 *LINK* Global Component-Based IT Architecture

LINK-Solution is supported by an ICT architecture that includes multi-computer systems for the several voltage levels, each having its Control Centre (CoCe). Bidirectional communication paths between characterise them as follows:

- HV_CoCe and MV_CoCe;
- MV_CoCe and LV_CoCe.

The LV_CoCe can communicate only with MV_CoCe and not with HV_CoCe, Fig. 2.62. The classification here is made arbitrarily. Depending on the utility organisation, the control centre may control MV and LV grids (MV-LV_CoCe), and so on (Fig. 2.63).

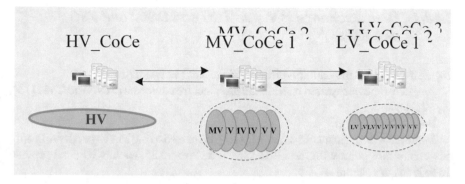

Fig. 2.63 *LINK* global component-based IT architecture

2.10 Data Privacy and Cybersecurity

Data privacy and cybersecurity are directly related to a large number of exchanged data.

> Massive **data exchanges** occur when many not owned devices are used to control and operate the system.

In the actual Smart Grids developments, this phenomenon occurs in two cases:

- TSO-Distributed generators and -customer plants [32] ; and.
- DSO-Customer plants.

The fundamental principle of the distributed *LINK*-Architecture prohibits access to all resources by default, allowing access only through well-defined interfaces. Thus, each Link is by definition modular and closed in itself, complying with the General Data Protection Regulation (GDPR). A distributed control system with limited communication between its smaller components is potentially more resistant to cyberattacks than centralised control systems that require a large amount of data exchange from a cybersecurity perspective.

> The distributed *LINK*-Architecture minimises data exchange by design, thereby facilitating cybersecurity, given the higher resilience of a distributed architecture.

2.10.1 Data Exchange Between TSO and DSOs and TSO and DGs

Figure 2.64 shows the data flow from DSO to TSO in two cases:

A—Actual power system operational architecture conforms to the Grid Code [13], Fig. 2.64a; and,

B—*LINK*-Solution.

There are *n* Significant Grid Users (SGU) connected to the MVG-part having only one connection point with the HVG. Based on Article 25, each SGU shall provide three kinds of schedules:

- The scheduled unavailability;
- The forecasted scheduled active power output at the connection point in the distribution grid; and.
- Any forecasted restriction in the reactive power control capability.

Article 29 describes two communication variants:

- Firstly, all schedules may be communicated by each SGU directly to the corresponding TSO and DSO, Fig. 2.64a; or,
- Secondly, they may be communicated via its DSO to the TSO, Fig. 2.64b.

Therefore, TSO will receive 3 *n* schedules in both variants of case A.

Figure 2.64c shows the data should be changed using the *LINK*-Architecture, case B. In this case, the SGU owners should exchange the data only with the Link operator where they are connected. Due to the enclosed nature of the Links, the TSO shouldn't get any information about the network users who are connected directly to

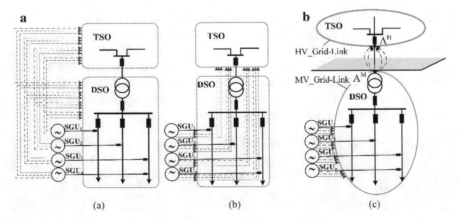

Fig. 2.64 Data flow from DSO to TSO: **a** Demand facilities communicate directly to the corresponding TSO and DSO; **b** Demand facilities communicate via the corresponding DSO to the TSO; **c** *LINK*-Solution

the distribution grid. That means they should communicate only with the DSO. The TSO will receive the required scheduled data from the DSO.

The exchanged data are the day ahead of scheduled active and reactive power, and the corresponding active and reactive power support $P_{Schedule}^{Day-ahead} \pm \Delta P$, $Q_{Schedule}^{Day-ahead} \pm \Delta Q$, that flow in the intersection point HV/MV; A^H/A^M. The number of the exchanged data is always four. As a result, the scheduled data amount that should be exchanged in the traditional architecture combined with the Grid Code increases continuously with the SGU number by $3 \cdot n$.

> In *LINK*-**Solution**, the number of exchanged schedules is independent of the SGU number. It remains constant at four.

2.10.2 Data Exchange Between DSO and Customer Plants

The massive integration of rooftop photovoltaic facilities causes upper voltage limit violations in low-voltage grids. To eliminate these voltage violations, the customer PV-inverters are upgraded with different local control strategies. Therefore, the LVG operation is intertwined with the operation of each LVG connected inverter, although the latter is the customers' property [28]. Figure 2.65 shows a schematic of the interaction between the DSO and the customer's PV-inverter.

The reactive power Q provided by each inverter depends on the feeder bus voltage ($U_{FeederBus}$) where the house and, as a consequence, the inverter is connected. All of the current solutions intended to prevent upper voltage limit violations cause new technical and social problems. PV inverters using $Q^{inv}(U)_{FeederBus}$(but also in the case using $\cos\varphi^{inv}(P^{inv})$) controls cause an excessive reactive power flow, thus increasing the grid losses and transformer loading considerably. In many cases, active power curtailments are necessary to ensure the quality and reliability of supply. Their needed

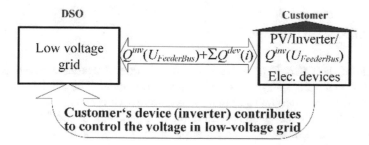

Fig. 2.65 Schematic of the interaction between the DSO and customer's PV-inverter

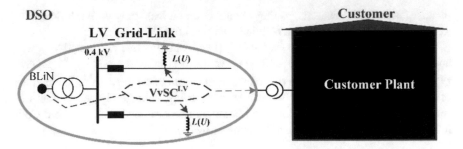

Fig. 2.66 Overview of the LV_Grid-Link using the concentrated $L(U)$ local control strategy

coordination provokes significant ICT challenges, Fig. 2.65, and their resolution is not yet foreseeable.

LINK-Solution postulates the use of the own device controls to maintain the required operation quality. DSO-owned inductive devices (inverter, or coils, or capacitors) are used to alleviate voltage violations in low voltage grids, Fig. 2.66. This control concept unbundles the operation of DSO- and customer-owned appliances, thus mitigating the aforementioned social issues.

> *LINK*-**Solution** postulates the primary use of the own reactive device controls to maintain the required voltage quality. It unbundles the operation of DSO- and customer-owned appliances, thus mitigating under others the ICT challenge.

Additionally, it brings significant technical advantages (see Chap. 4). The locally controlled $L(U)$ are set at the end of the violated feeders. $L(U)$ control strategy needs very little data if coordination is required. Figure 2.67 shows schematically a LVG and the required data flow for coordination in two cases: by using the customer owned $Q(U)$ control, Fig. 2.67a and the DSO-owned $L(U)$-control device, Fig. 2.67b. The difference in the amount of data exchange required is clear.

2.11 *LINK*-Economics

The basic of the *LINK*-Economics is the functioning of available advanced technologies of management systems and optimal investment. The available advanced technologies have undergone an intensive scientific review and are demonstrated by the industry for both energy and distribution management systems. Therefore, there are established and reliable advanced applications that work on some upgrades to the specifics of Links. In other words, the accumulated knowledge and the available technologies for the operation of power systems are helpful to *LINK*-Solution. However, the latter has some unique aspects that require innovations. In general,

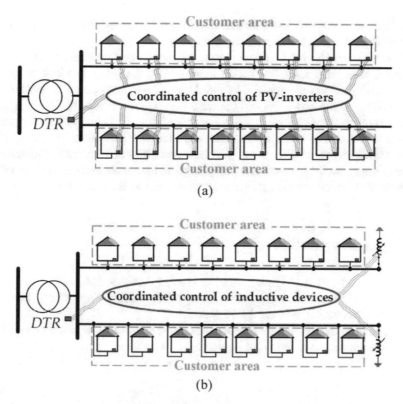

Fig. 2.67 Schematic LVG and the required data flow for coordination of **a** customer-owned PV-inverters in case of $Q(U)$ control; and **b** DSO-owned $L(U)$-control device suggested by *LINK*-Solution

the *LINK*-Solution differs significantly from the traditional one because it consists of Links that work independently together. New online and real-time applications should be designed and developed to ensure reliable and resilient operation through the dynamic optimisation in each Link. The large-scale integration of DERs and various advanced technologies enabled by *LINK*-Solution will reduce the capital expenditures of the future power systems, Smart Grids. For example, the large-scale integration of DERs into the existing radial structures of medium and low voltage grids and customer plants will defer or wholly avoid the cost of system extension that would otherwise be necessary to cope with the growing load. The dynamic joint optimisation of demand and supply is a new opportunity in power systems economics. In traditional power systems, load control is usually treated during the planning and off-line analyses as demand-side management or associated with differentiated tariffs or contracts.

ECs realised under the *LINK*-Solution behave differently. The marginal costs of own power production will be taken over at any time by the EC participants. In addition, They will have the opportunity to consider the marginal costs of energy provision at all times together with the related costs for energy efficiency investments and demand reduction based on demand response.

Power systems have traditionally been developed and operated to provide all customers with electricity that meets more or less the same standard of quality and reliability of power supply. *LINK*-Solution can control power reliability and quality closer to the end-users and optimise these characteristics for the respective loads. Therefore, heterogeneous standards of quality and reliability of power supply are possible. Based on and through combined economic and environmental studies, optimal investments in DERs or grid reinforcements may be identified. With the new market design, each Link operator participates in the whole market, thus enabling power trading beyond small entities' substation and ancillary services.

2.12　Implementation

A dynamic and efficient implementation plan of *LINK*-Solution is crucial for a smooth transition and successful realisation of Smart Grids. Figure 2.68 shows the implementation structure of the *LINK*-Solution.

According to the emerging climate needs, three significant steps are required for an effective, successful, and rapid implementation on a large scale:

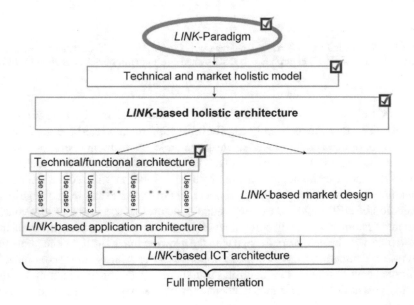

Fig. 2.68 The implementation structure of *LINK*-Solution

1. **The design of applications architecture** relies on the detailed specification of the use cases—the latter need to be defined from the operational processes of Smart Grids.

2. **The *LINK*-Market structure's design** harmonizes with the physical laws of power systems and allows the active participation of all stakeholders regardless of size and type;

3. **The design of the *LINK*-ICT architecture** derives from the two steps above.

2.12.1 Relation Between Automation and Digitalisation

The **automation** concept is ancient; it is introduced around 270 BC. With the emergence of computer technologies in the mid-twentieth century, **digitalisation** has become an inseparable part of automation. In contrast, **digitalisation** is the result of Internet Technologies developed at the beginning of the twenty-first century.

Different parts of power systems (for example, generation) were an automation object from the beginning. Later, the introduction of SCADA and various management systems made it possible to increase the level of automation. Nowadays, the integration of distributed generation and the highly volatile nature of the new renewable energy sources (wind, solar) require a high automation degree of power systems and customers.

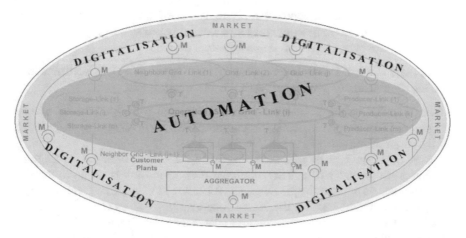

Fig. 2.69 Overview of the application areas of automation and digitalisation techniques in the *LINK*-Architecture

> **Automation** is the core of the *LINK*-based holistic architecture and a prerequisite for successful market design and the effective implementation of digitalisation Fig. 2.69.

Automation shapes the physical behaviour of power systems, followed and respected by the human-made market rules. Digitalisation is necessary to enable the active participation of customers, electricity producers, and storage operators into the market.

2.12.2 Links Arrangement

LINK-Solution reorganises the grid, electricity production, energy storage facilities, and consumers by dividing the whole system into clearly defined units, "Links" each with its control system and well-defined interfaces to their neighbouring Links and the market. Grid-Links may be arranged throughout the entire grid from the HV to the grids CPs in different sizes defined by the secondary control areas. The following are discussed some typical Grid-Link arrangements for both power system approaches, European and North American, and a Smart City district.

2.12.2.1 European and North American Power System Practices

Electric power distribution system designs and practices have evolved in different directions worldwide due to various factors. Most can be traced back to either the 'European' or 'North American' approach [6]. Figure 2.70 shows different Grid-Link arrangements for different approaches: European and North American approaches by maintaining the existing structures and a new proposed structure.

There do exist two types of systems operators actually:

1. The TSO in Europe and the Independent Systems Operator (ISO) in North America, who are responsible for the operation of the transmission grid;

2. The DSOs for the North American type of distribution grid are accountable for the primary and secondary grid, while the European ones are in many cases responsible for the sub-transmission and the MV and LV grid.

In the European type of grid, the HV^T_Grid-Link may be arranged on the transmission grid operated by the TSO, while an HV^S_Grid-Link is set up in the sub-transmission grid operated by the DSO, Fig. 2.70a. On MVGs and LVGs are arranged MV_ and LV_Grid-Links. In the North American grid type, the HV^T_Grid-Link may be arranged on the transmission grid operated by the ISO, while the MV_ and LV_Grid-Link types are set up on distribution grids operated by the DSO, Fig. 2.70b.

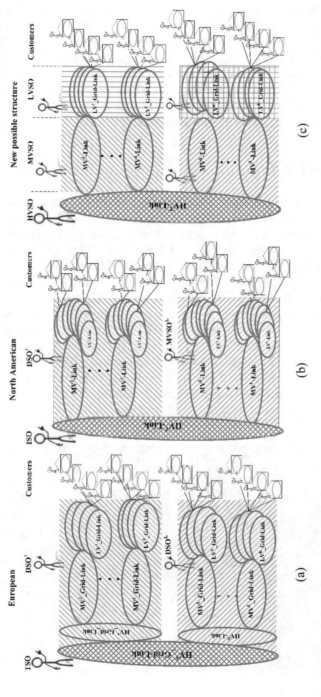

Fig. 2.70 Different Grid-Link arrangements for different approaches: **a** European; **b** North American; **c** New possible structure

Figure 2.70c shows a new possible Link-Grid structure, which provides the unbundling of the distribution operation on medium/primary and low/secondary voltage grid, i.e. MV_ and LV_Grid-Link operation. In this case, the power grid will be operated from three types of operators: HV-, MV- and LV System Operators (HVSO, MVSO, and LVSO) responsible for the operation of the HV_, MV_ and LV_Grid-Link, respectively. *LINK*-Solution stipulates an operator for each Link or Link-bundle type. CP_Grid-Link is a special one because the owner is the customer who usually isn't aware of it. Therefore, the reliable active operation of this Grid-Link type realises only in the case of its full automation.

2.12.2.2 Smart Cities

Smart Cities *are where decarbonisation* strategies for energy, transport, buildings, and even industry and agriculture coexist and intersect [17]. *LINK*-Solution supports the decarbonisation of all economic sectors by enabling Energy Systems Integration in the context of smart cities (see Sects. 3.1 and 3.2). The flexible setting of the Grid-Link size allows the easy application of *LINK*-Solution to the Smart City grid. Figure 2.71 shows Links arrangement in a city district. An LV subsystem with a distribution transformer and two feeders connects a residential building, a district PV facility, a district garage with e-cars and many single-family houses. On the residential building is installed a common roof PV system for all residents, while on the basement is a garage with e-cars. In the one family houses are also installed rooftop PV facilities, while on the basement of each house is a garage with an e-car. But these are not shown in detail in the *LINK*-architecture shown in Fig. 2.72. It depicts the technical/functional architecture of the district. Two LV_Grid-Links are identified: the one includes the distribution transformer and the feeders, the other the

Fig. 2.71 Links arrangement in a city district

grid within the residential building: The corresponding secondary controls are set up on the LV grids. CP_Grid-Links are put on each flat and one family house. On the PV facilities and charging station of e-cars are set Producer- and Storage-Links.

Figure 2.73 shows the resilient connection between the different Links of a Smart City district. The dynamically controlled active and reactive power flows between various Links allow their robust and resilient interaction.

Each Link acts as a black box that exchanges a limited amount of data with other neighbouring Links, thus guaranteeing their data privacy Fig. 2.74.

LINK-Solution enables Smart Grids in cities while maintaining **data privacy** by avoiding citizen surveillance.

As electricity is the most critical infrastructure of our society, **IoT technologies** are strictly separated from the *LINK*-Technology.

Of course, each customer has the right to decide for himself about the use of IoT technologies.

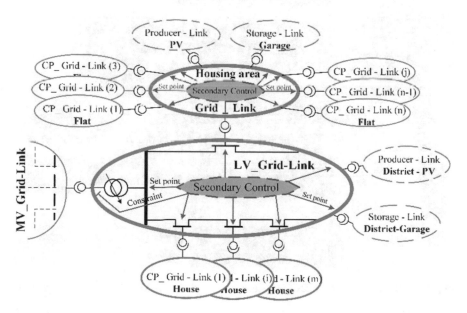

Fig. 2.72 Technical/functional architecture of a Smart City district

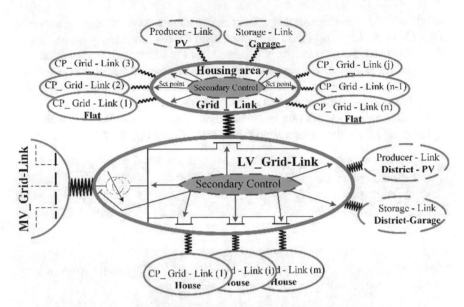

Fig. 2.73 The resilient connection between different Links of a Smart City district

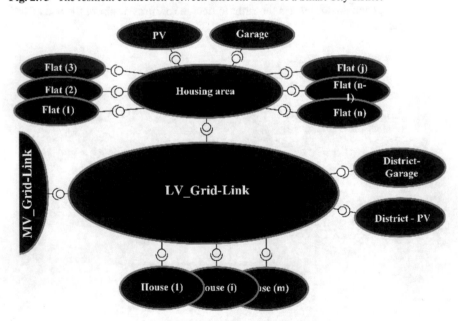

Fig. 2.74 Data privacy of each Link is guaranteed

2.12.3 Transition Period

The upgrade of the power system architecture is compelling but won't be built in a day—or a decade. Consequently, the upgrade process will be accompanied by a transition period with a hybrid architecture. During all this time, the upgrade will be done stepwise to ensure a secure, reliable, and feasible operation of the entire power system. The essential upgrade steps are presented in the following using two methods:

- **Top-down** method: HVG and the power plants that are feeding it build the power system's backbone. Consequently, the consolidation of the Volt/var loop in medium voltage level with well-defined constraints on the boundary with the high voltage level has the highest priority. After that, the HV- and the LV levels may be consolidated simultaneously. The consolidation of the loops concerning active power/frequency should follow the Volt/var ones. The developments in the CP level will follow, and it is expected that they will last longer;
- **Bottom-up** method: The development starts at the CP level by employing energy communities (see Sect. 3.3). Tackling voltage challenges and the uncontrolled reactive power in the superordinated grids is the first challenge to be managed in the vertical axis (see Sect. 4), followed by the active power balance. Development on the transmission grids follows.

Appendix

A.2.1 Load

Depending on the topological position of the delivery point, the load can represent:

- The flow of individual devices such as motors, lighting, and so on, Fig. 2.75;
- The aggregated flow of hundreds or thousands of individual component devices such as motors, lighting, and so on, Fig. 2.76;
- The aggregated flow of individual devices combined with electrical appliances such as sub-transmission lines, distribution feeders, customer plants wiring, reactive power devices, transformers, etc., Fig. 2.77.

The load can also represent the total amount of power that should be delivered from the electricity producer. In this case, it is called "Power System Load," and it includes the electricity supplied to all users connected to the distribution network of a system, and also the power provided to compensate for losses in all parts of the network (transformers, converters, and transmission lines, and so on).

Fig. 2.75 Load represents the flow of individual devices such as motors, lighting, etc.: **a** Load modelling; **b** Individual device

P_{Load}

Q_{Load}

Dev.

P_{Load}

Q_{Load}

(a)   (b)

Fig. 2.76 Load represents the aggregated flow of hundreds or thousands of individual component devices such as motors, lighting, etc.: **a** Load modelling; **b** Many individual devices

$$P_{Load} = \sum_1^n P_{Load}(i)$$

Dev.

$$Q_{Load} = \sum_1^n Q_{Load}(i)$$

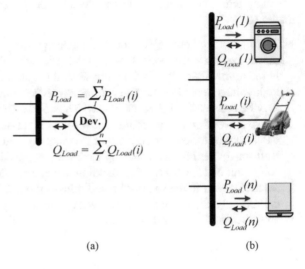

$P_{Load}(1)$

$Q_{Load}(1)$

$P_{Load}(i)$

$Q_{Load}(i)$

$P_{Load}(n)$

$Q_{Load}(n)$

(a)   (b)

Fig. 2.77 Load represents the aggregated flow of individual devices combined with electrical appliances such as sub-transmission lines, distribution feeders, customer plants, etc.: **a** Load modelling; **b** Grid part

$P_{Load} = P_{Tr.}$

$Q_{Load} = Q_{Tr.}$

#

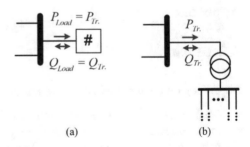

$P_{Tr.}$

$Q_{Tr.}$

(a)   (b)

A.2.2 Controls

A.2.2.1 Popular Control Strategies in Power Systems

The most popular control strategies in power systems are local controls in open and closed loops. When the secondary control loops are set, the local controls in closed loops are usually called primary controls.

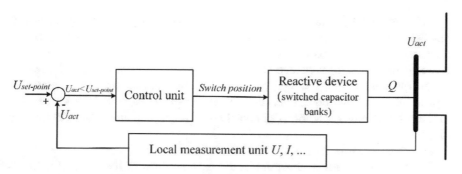

Fig. 2.78 Open action path of the Local Control of switched capacitor banks

Local control

Local Control refers to control actions that are carried out locally without considering the holistic real-time behaviour of the relevant grid part. Its action path may be realised in open- or closed-loop:

- Open-loop path—The input variable usually differs from the output one; the output variables are influenced by the input variables but **do not act on themselves continuously and again** via the same input variables. Figure 2.78 shows the open-loop action path of a switched-capacitor bank, where the output variable is always reactive power. In contrast, the input variable may be voltage, current, time, and so on.
- Closed-loop path—In this case, **the controlled variable continuously influences itself**. The deviation of the actual measured value from set-point results in a signal, which affects the valves or frequency, excitation current or reactive power, transformer steps, and so on in such a way that the desired power is delivered or the desired voltage is reached. Figure 2.79 shows the closed-loop action path of the Local Control of OLTC. It keeps the voltage to the Uset-point.

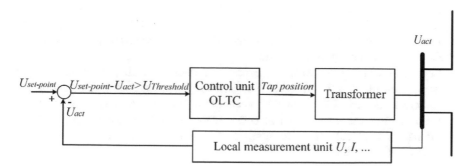

Fig. 2.79 Closed action path of the Local Control of switched capacitor banks

LC automatically adjusts the active/reactive power contributions, tap and switch positions, etc., of the corresponding control device based on local measurements or time schedules [17, 58, 65]. They usually maintain a power system parameter, which is locally measured or calculated based on local measurements, equal to the desired value. The fixed control settings are calculated based on offline system analysis for typical operating conditions. Local controls are simple, reliable, and quickly respond to changing operating conditions without the need for a communication infrastructure [53, 59, 74].

Secondary control

SC in power systems is quite popular in the case of load frequency control. LFC's significant purposes are maintaining the operation area's frequency and keeping power exchange in the tie lines conform to the schedules. PC's objective is to maintain a balance between generation and consumption (demand) within the synchronous area. SC maintains a balance between generation and consumption (demand) within each control area and the synchronous area's system frequency. Tertiary control is primarily used to free up the SC reserves in a balanced system situation by considering the economic dispatch [14].

A.2.2.2 Control Set Used in LINK-Solution

A control set is used in the *LINK*-Solution that consists of a Direct, Primary, and Secondary Control loop, Fig. 2.80.

- Primary Control (PC) refers to control actions done locally in a closed-loop: the input and output variables are the same. The output- or control-variable is locally measured and continuously compared with the reference variable, the setpoint calculated by secondary control. The deviation from the setpoint leads to a signal that influences the valves or frequency, excitation current or reactive power, transformer steps, etc., in a primary-controlled power plant, transformer, etc., so that the desired power is delivered or the desired voltage is reached.
- Direct Control (DiC) refers to control actions performed in an open-loop, taking into account the real-time holistic behaviour of the grid part he belongs. Secondary Control calculates its control action.
- Secondary Control (SC) refers to control variables calculated based on a control area's current state. It fulfils a predefined objective function by respecting the static, i.e., electrical appliances' constraints (PQ diagrams of generators, transformer rating, etc.), and dynamic conditions dictated by neighbouring areas. At the same time, it calculates and sends the setpoints to PCs and the input variables to DiC acting on its control area.

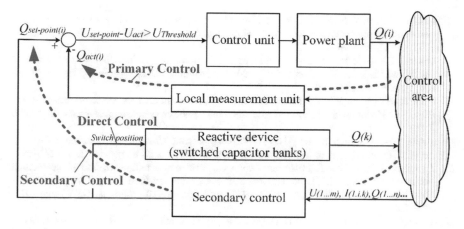

Fig. 2.80 Overview of the control set used in *LINK*-Solution

References

1. Albadi M, El-Saadany E (2008) A summary of demand response in electricity markets. Department of Electrical and Computer Engineering, University of Waterloo, vol 78/11, pp 1989–1996
2. Amin M (2013) Energy: The smart-grid solution. Nature 499:145–147. https://doi.org/10.1038/499145a
3. Bollen MHJ, Sannino A (2005) Voltage control with inverter-based distributed generation. IEEE Trans Power Delivery 20:519–520. https://doi.org/10.1109/TPWRD.2004.834679
4. Bose A (2010) Smart transmission grid applications and their supporting infrastructure. IEEE Trans Smart Grid 1:11–19. https://doi.org/10.1109/TSG.2010.2044899
5. Boyer SA (2010) SCADA supervisory control and data acquisition. ISA—International Society of Automation, p 179. ISBN 978-1-936007-09-7
6. Carr J, McCall LV (1992) Divergent evolution and resulting characteristics among the world's distribution systems. IEEE Trans Power Delivery 7(3):1602–1609
7. Clarke GR, Reynders D, Wright E (2004) Practical modern SCADA protocols: DNP3, 60870.5 and related systems Newnes. ISBN 0750657995, 9780750657990
8. Commission Regulation EU (2017)/1485 of 2 August 2017—establishing a guideline on electricity transmission system operation. Official J Eur Union. https://eur-lex.europa.eu/legal-content/EN/TXT/PDF/?uri=CELEX:32017R1485&from=EN. Accessed 10 Oct 2018
9. Degaudenzi ME, Arizmendi M (2000) Wavelet based fractal analysis of electrical power demand. Fractals 8:239–245. https://doi.org/10.1142/S0218348X0000024X
10. Dzafic I, Gilles M, Jabr RA, Pal BC, Henselmeyer S (2013) Real time estimation of loads in radial and unsymmetrical three-phase distribution networks. IEEE Trans Power Syst 28(4):4839–4848
11. D'hulst R, Fernandez JM, Rikos E, Kolodziej D, Heussen K, Geibelk D, Temiz A, Caerts C (2015) Voltage and frequency control for future power systems: the ELECTRA IRP proposal. In: International Symposium on Smart Electric Distribution Systems and Technologies (EDST), Vienna, 8–11 September, pp 245–250. https://doi.org/10.1109/SEDST.2015.7315215
12. E-control (2013) The Austrian electricity market, pp 1–20. https://www.e-control.at/documents/1785851/1811528/Strommarktmodell_%C3%96sterreich_030413_en.pdf/f007c31e-b6b9-48a1-83a1-53212aae5365?t=1413907916729

13. ENTSO-E (2013) Network Code on Operational Security, 24 September. https://www.ent soe.eu/major-projects/network-code-development/operational-security/Pages/default.aspx. Accessed 12 September 2016

14. ENTSO-E (2015) P1—Policy 1: load-frequency control and performance. https://eepublicd ownloads.entsoe.eu/clean-documents/pre2015/publications/entsoe/Operation_Handbook/Pol icy_1_final.pdf. Accessed 11 May 2021

15. ETIP SNET (2019) White Paper Holistic architectures for future power systems. https://www. etip-snet.eu/white-paper-holistic-architectures-future-power-systems. Accessed 12 March 2020

16. Etherden N, Vyatkin V, Bollen HJ (2015) Virtual power plant for grid services using IEC 61850. In: IEEE transaction on industrial informatics, pp 1–11. https://doi.org/10.1109/TII.2414354

17. Farivar M, Zho X, Chen L (2015) Local voltage control in distribution systems: an incremental control algorithm. In: IEEE international conference on Smart Grid Communications (Smart-GridComm), Miami, FL, USA, pp 732–737. https://doi.org/10.1109/SmartGridComm.2015. 7436388

18. Florea G, Chenaru O, Popescu D, Dobrescu R (2015) A fractal model for Power Smart Grids. In: IEEE 20th international conference on control systems and science, Bucharest, 27–29 May, pp 572–577

19. Frankhauser P, Tannier C, Vuidel G (2018) Fractalyse 2.4. Research Centre ThéMA. http:// www.fractalyse.org/en-home.html. Accessed 13 Oct 2019

20. Gellings CW, Parmenter KE (2016) Demand-side management. In: Energy efficiency and renewable energy handbook, 2nd edn

21. Gupta A (2014) The world's longest power transmission lines. In: Power tech-nology. https://www.power-technology.com/features/featurethe-worlds-longest-power-transm ission-lines-4167964. Accessed 09 June 2016

22. Herzum P, Oliver S (2000) Business components factory: a comprehensive overview of component-based development for the enterprise. Wiley. ISBN 978-0-471-32760-8

23. Hou H, Tang A, Fang H, Yang X, Dong Z (2015) Electric power network fractal and its relationship with power system fault. Tehnicki Vjesnik 22:623–628. https://doi.org/10.17559/ TV-20150427180553

24. Huang SJ, Lin JM (2003) Application of box counting method-based fractal geometry tech-nique for disturbance detection in power systems. In: IEEE power engineering society general meeting, Toronto, 13–17 July, vol 3, pp 1604–1608. https://doi.org/10.1109/PES.2003.1267395

25. IEEE Task Force on Load Representation for Dynamic Performance (1993) Load representation for dynamic performance analyses (of power systems). IEEE Trans Power Syst 8: 472–482. https://doi.org/10.1109/59.260837

26. Ilo A (2013) The energy supply chain net. Energy Power Eng 5:384–390. https://doi.org/10. 4236/epe.2013.55040

27. Ilo A (2019) Design of the Smart Grid architecture according to fractal principles and the basics of corresponding market structure. Energies 12:4153. https://doi.org/10.3390/en12214153

28. Ilo A, Schultis DL (2019) Low-voltage grid behaviour in the presence of concentrated var-sinks and var-compensated customers. Electric Power Syst Res 171:54–65. https://doi.org/10.1016/ j.epsr.2019.01.031

29. Ilo A (2016/1) "Link"—the smart grid paradigm for a secure decentralised operation architecture. Electric Power Syst Res 131:116–125. https://doi.org/10.1016/j.epsr.2015.10.001

30. Ilo A (2016/2) Effects of the reactive power injection on the grid—the rise of the Volt/var interaction chain. Smart Grid Renew Energy 7:217–232. https://doi.org/10.4236/sgre.2016. 77017

31. Ilo A (2016/3) Minimization of exchanged data on the TSO-DSO cross border by applica-tion of a new operation architecture. In: CIGRE International Colloquium, Philadelphia, 2–3 November

32. Ilo A (2017/1) Are the current smart grid concepts likely to offer a complete Smart Grid solution? Smart Grid Renew Energy J 7:252–263. https://doi.org/10.4236/sgre.2017.87017

33. Ilo A (2017/2) Demand response process in context of the unified LINK-based architecture. In: Bessède J-L (ed) Eco-design in electrical engineering eco-friendly methodologies, solutions and example for application to electrical engineering. Springer, Heidelberg, ISBN 978-3-319-58171-2

34. Ilo A, Gawlik W, Schaffer W, Eichler R (2015) Uncontrolled reactive power flow due to local control of distributed generators. In: 23th international conference on electricity distribution, CIRED, Lyon, 15–18 June, pp 1–5

35. Ilo A, Prata R, Iliceto A, Strbac G (2019) Embedding of energy communities in the unified LINK-based Holistic Architecture. CIRED, Madrid, 3–6 June, pp 1–5

36. Ilo A, Reischböck M, Jaklin F, Kirschner W (2005) DMS impact on the technical and economic performance of the distribution systems. In: 18th international conference and exhibition on electricity distribution, CIRED, Turin, pp 1–5

37. Ilo A, Schaffer W, Rieder T, Dzafic I (2012) Dynamische Optimierung der Verteilnetze—closed loop Betriebergebnisse. VDE Kongress, Stuttgart, 5–6 November, pp 1–6. https://www.vde-verlag.de/proceedings-de/453446029.html. Accessed 19 Mai 2015

38. Ilo A, Schultis DL, Schirmer C (2018) Effectiveness of distributed vs. Concentrated Volt/Var local control strategies in low-voltage grids. Appl Sci 8:1382. https://doi.org/10.3390/app808 1382

39. Kießling A (2013) Beiträge von moma zur Transformation des Energiesystems für Nach-haltigkeit, Beteiligung, Regionalität und Verbundheit. Modellstadt Mannheim (moma). https://www.ifeu.de/energie/pdf/moma_Abschlussbericht_ak_V10_1_public.pdf. Accessed

40. Kirshner D, Giorsetto P (1984) Statistical test of energy saving due to voltage reduction. IEEE Trans Power Appar Syst 6:1205–1210

41. Klima Energie fonds (2012) ZUQDE-project final report, pp 1–111. https://www.energieforsc hung.at/assets/project/final-report/ZUQDE.pdf. Accessed 11 Mai 2021

42. Le TTM, Retière N (2014) Exploring the scale-invariant structure of Smart Grids. IEEE Syst J 11:1612–1621. https://doi.org/10.1109/JSYST.2014.2359052

43. Lefebvre S, Gaba G, Ba AO, Asber D, Ricard A, Perreault C, Chartrand D (2008) Measuring the efficiency of voltage reduction at Hydro-Québec distribution. In: 2008 IEEE power and energy society general meeting-conversion and delivery of electrical energy in the 21st Century, Pittsburgh, 20–24 July, pp 1–7

44. Lo KL, Peng LJ, Macqeen JF, Ekwue AO, Cheng DTY (1997) An extended Ward equivalent approach for power system security assessment. Electric Power Syst Res 42:181–188

45. Ma J, Wang Z (2007) Application of Grille fractal in identification of current transformer saturation. Power System Technol 31:84–88

46. Maier MW, Rechtin E (2009) The art of systems architecting. CRC Press, Taylor & Francis Group, Boca Raton, ISBN 9781420079135

47. Mamishev AV, Russell BD, Benner CL (1995) Analysis of high impedance faults using fractal techniques. IEEE Trans Power Syst 11:435–440. https://doi.org/10.1109/59.486130

48. Mandelbrot BB (1983) The fractal geometry of nature. W. H. Freeman and Company, Dallas, ISBN 0716711869

49. Marris E (2008) Upgrading the grid. Nature 454:570–573. https://doi.org/10.1038/454570a

50. Mayer M et al (2009) The art of system architecting. CRC Press

51. Miller C, Martin M, Pinney D, Walker G (2014) Achieving a resilient and agile grid. The National Rural Electric Cooperative Association, Arlington, pp 1–78. http://www.electric.coop/wp-con-tent/uploads/2016/07/Achieving_a_Resilient_and_Agile_Grid.pdf. Accessed 11 May 2021

52. Moslehi K, Kumar R (2010) A reliability perspective of the Smart Grid. IEEE Trans Smart Grid 1:57–64. https://doi.org/10.1109/TSG.2010.2046346

53. Nowak S, Wang L, Metcalfe MS (2020) Two-level centralized and local voltage control in distribution systems mitigating effects of highly intermittent renewable generation. Int J Electrical Power Energy Syst 119:105858

54. Ortjohann E, Wirasanti P, Schmelter A, Saffour H, Hoppe M, Morton D (2013) Cluster fractal model - A flexible network model for future power systems. In: International conference on clean electrical power, Alghero, 11–13 June, pp 293–297. https://doi.org/10.1109/ICCEP.2013.6587004

55. Preiss RF, Warnock VJ (1978) Impact of voltage reduction on energy and demand. IEEE Trans Power Apparatus Syst 97:1665–1671. https://doi.org/10.1109/TPAS.1978.354658

56. Raab AF, Ferdowsi M, Karfopoulos E, Unda I, Skarvelis-Kazakos S, Papadopoulos P, Abbasi E, Cipcigan L, Jenkins N, Hatziargyriou N, Strunz K (2016) Virtual power plant control concepts with electric vehicles. In: 16th international conference Intelligent System Application to Power Systems (ISAP), pp 1–6.

57. Roytelman I, Shahidehpour SM (1993) State estimation for electric power distribution systems in quasi-real-time conditions. IEEE Trans Power Delivery 8(4):2009–2015

58. Roytelman I, Ganesan V (1999) Modeling of local controllers in distribution network applications. In: Proceedings of the 21st international conference on power industry computer applications. Connecting Utilities. PICA 99. To the Millennium and Beyond (Cat. No.99CH36351), Santa Clara, CA, USA, pp 161–166. https://doi.org/10.1109/PICA.1999.779399.

59. Roytelman I, Ganesan V (2000) Coordinated local and centralized control in distribution management systems. IEEE Trans Power Delivery 15/2(2):718–724. https://doi.org/10.1109/61.853010

60. Sanduleac M, Pop R, Borza P (2012) Smart Grid patterns of fractality. Energetica 60:188–191

61. Schultis DL, Ilo A, Schirmer C (2019) Overall performance evaluation of reactive power control strategies in low voltage grids with high prosumer share. Electric Power Syst Res 168:336–349. https://doi.org/10.1016/j.epsr.2018.12.015

62. Schweppe FC, Wildes J (1970) Power system static-state estimation. IEEE Trans Power Apparatus Syst PAS-89/1:120–125. https://doi.org/10.1109/TPAS.1970.292678

63. Schweppe FC, Rom DB (1970) Power system static-state estimation, Part II: Approximate Model. IEEE Trans Power Apparatus Syst PAS-89/1:125–130. https://doi.org/10.1109/TPAS.1970.292679

64. Strbac G (2008) Demand side management: benefits and challenges. Energy Policy 36(12):4419–4426

65. Sun H et al (2019) Review of challenges and research opportunities for voltage control in Smart Grids. IEEE Trans Power Syst 34(4):2790–2801. https://doi.org/10.1109/TPWRS.2019.2897948

66. Taft J, Martini PD, Geiger R (2014) Ultra large-scale power system control and coordination architecture—a strategic framework for integrating advanced grid functionality. U.S. Department of Energy, Washington DC. https://gridarchitecture.pnnl.gov/media/white-papers/ULS%20Grid%20Control%20v3.pdf. Accessed 11 May 2021

67. Thorp JS, Naqavi SA (1989) Load flow fractals. In: 28th IEEE conference on decision and control, Tampa, 13–15 December, vol 2, pp 1822–1827. https://doi.org/10.1109/CDC.1989.70472

68. Thorp JS, Naqavi SA, Chiang HD (1990) More load flow fractals. In: 29th IEEE conference on decision and control, Honolulu, 5–7 December. IEEE Xplore, vol 6, pp 3028–3030. https://doi.org/10.1109/CDC.1990.203339

69. Vaahedi E (2014) Practical power system operation. Wiley, New Jersey

70. Viswanadh SR, Raju VB (2016) Application of fractals to power system networks. IJSETR, vol 5, pp 2547–2553. http://ijsetr.org/wp-content/uploads/2016/08/IJSETR-VOL-5-ISSUE-7-2547-2553.pdf. Accessed 24 Mai 2021

71. Wang F, Li L, Li C, Wu Q, Cao Y, Zhou B, Fang B (2018) Fractal characteristics analysis of blackouts in interconnected power grid. IEEE Power Eng Lett 33:1085–1086. https://doi.org/10.1109/TPWRS.2017.2704901

72. Wang Z, Ma JA (2007) Novel method to identify inrush current based on Grille fractal. Power Syst Technol 31:63–68

73. Wolak FA (2003) Diagnosing the California electricity crisis. Electr J 16:11–37

74. Zhou X, Farivar N, Liu Z, Chen L, Low SH (2021) Reverse and forward engineering of local voltage control in distribution networks. IEEE Trans Autom Control 66(3):1116–1128. https://doi.org/10.1109/TAC.2020.2994184

Chapter 3
Energy Systems Integration*

> Avoiding climate breakdown will require cathedral thinking. We
> must lay the foundation while we may not know exactly how to
> build the ceiling.
> —Greta Thunberg

Environment protection policies and overall climate commitments require the
decarbonisation of all sectors of the economy (European Commission [7]). The
latter has evolved from individual sectors with little or no dependencies into an
intertwined and complex structure. Therefore, the decarbonisation of the economy
in all sectors is one of the most significant challenges of this century. The integration
of Energy Systems of various economic sectors is considered the most suitable way
to decarbonise them and reduce CO_2 emissions.

This chapter deals with the relationship of *LINK*-Solution to the Integration of
Energy Systems, Sector Coupling, and Energy Communities.

3.1 *LINK*-Solution Relationship to Energy Systems Integration

The main goal of Energy Systems Integration (ESI) is the decarbonisation of the
economy (European commission [8]).

> By definition, **Energy Systems Integration** is the process of coordinating
> the operation and planning of energy systems across multiple pathways and
> geographical scales to deliver reliable, cost-effective energy services with
> minimal impact on the environment [16].

---

*Author: Albana Ilo

© Springer Nature Switzerland AG 2022
A. Ilo and D.-L. Schultis, *A Holistic Solution for Smart Grids
based on LINK–Paradigm*, Power Systems,
https://doi.org/10.1007/978-3-030-81530-1_3

ESI covers the entire economy of a country. The economy is divided into sectors to determine the proportion of a population involved in various activities [17]. Although many economic models divide the economy into only three sectors, others divide it into four or five. Our discussion focuses on the three first sectors: primary and tertiary (see Appendix A.3.1). The last two sectors: Quaternary- and Quinary Sector, are closely linked with the tertiary sector services, which is why they can also be grouped into it. Additionally, both last sectors are not directly associated with energy resources.

ESI is categorised into three opportunity areas: Streamline, Synergise, and Empower Fig. 3.1. This categorisation helps to identify how various ESI approaches can offer solutions to problems related to climate change.

Streamline refers to improvements made within the existing energy system of different sectors by restructuring, reorganising, and modernising current energy systems through institutional levers (i.e., policies, regulations, and markets) or investment in infrastructure. *LINK*-Solution modernises the electricity or power system in such a way that ESI can be realised.

Synergise describes ESI solutions that connect energy systems.

- of various vectors within the second sector (e.g. Cross different Vectors of the energy utilities such as electricity, heat, gas, and so on);

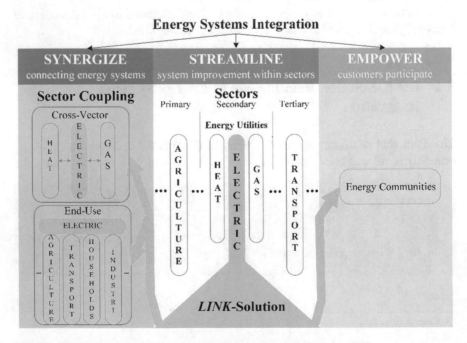

Fig. 3.1 *LINK*-solution in context of energy systems integration based on [16]

> By definition, **Cross-Vector Sector Coupling** involves integrating different energy infrastructures and vectors, particularly electricity, heat, and gas.

- between the electricity and End-Users (e.g. electricity and transport, agriculture, industry, households, and so on) to benefit efficiency and performance [18].

> By definition, **End-Use Sector Coupling** involves integrating different infrastructures within the customer plants such as electricity or power, heat and cooling infrastructures, and the modernisation of energy demand while reinforcing the interaction between electricity supply and end-use, such as electricity or power to chemistry and so on.

LINK-Solution describes and enables the Cross-Vector and End-Use Sector Coupling.

Empower refers to ESI actions that include the customer, whether through their investment decisions, active participation, or decisions to shift energy modes (e.g. through Energy Communities). *LINK*-Solution enables the creation and operation of Energy Communities.

3.2 Sector Coupling

Sector Coupling contributes to the cost-efficient decarbonisation of the economy and effective use of the existing infrastructures and renewable resources by using synergies and links between different sectors of the economy. Many studies have been done in Sectors Coupling, which has focused on the efficiency of its implementation [2, 9, 19]. The chemical energy carriers—like hydrogen and methane—produced via P2X processes—have the potential for large-scale, long-term (seasonal) energy storage capacity [3]. The latter may distinctly increase the grid flexibility. All of these are individual independent studies that have shown the potential of Sector Coupling. In the following, we focus on embedding Sector Coupling in the *LINK*-Architecture to enable its implementation on a large scale.

3.2.1 Sector Coupling Embedment in LINK-Architecture

In a landscape with integrated energy systems, fuel, thermal, water, and transport systems are systematically planned, designed, and operated as flexible "virtual

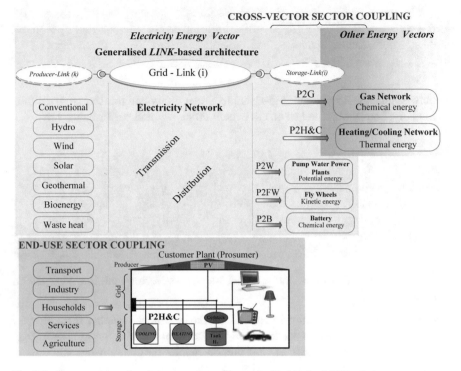

Fig. 3.2 Cross-vector and end-use sector coupling embedded in the *LINK*-solution

storage" resources for the power grid and vice versa [16]. Since Storage is defined as a unique and independent main element of the *LINK*-Architecture, the embedding of the Sector Coupling in it is obvious.

Figure 3.2 shows the Cross-Vector and End-Use Sector Coupling embedded in the *LINK*-Architecture. Energy vector areas are presented in different colours. The lime-green area presents the electricity energy vector, while the apricot area presents other vectors such as gas, heating and cooling, and so on. Its main elements outline the generalised *LINK*-Architecture (see Sect. 2.3.3.3):

- **Producers** include all available plants regardless of size and technology, such as conventional, hydro, wind, solar, geothermal, bioenergy, waste heating, etc.
- **Grids** include transmission (very high and high voltage level) and distribution (medium and low voltage levels); and
- **Storages** include all available facilities regardless of size and technology. It comprises the traditional storage facilities and the "virtual storage" resources.

The conventional storage facilities are part of the electricity energy vector. They include pumped hydroelectricity storages storing electricity in potential energy, flywheels storing electricity in kinetic energy, Batteries storing electricity in chemical energy, etc. The "virtual storages" are the result of the Cross-Vector Coupling exploiting the Power to Gas (P2G), Power to Heating and Cooling (P2H&C)

Table 3.1 Storage categorisation from the perspective of the grid

Storage-Link category	Description
Cat. A	The stored energy is injected at the charging point of the grid
Cat. B	The stored energy is not injected back at the charging point on the grid
Cat. C	The stored energy reduces the electricity consumption at the charging point in the near future

processes, and so on. These processes couple the infrastructures of electricity and gas utilities, electricity, and heating and cooling utilities. End-Use Sector Coupling happens in customer plant levels. The three main elements of the *LINK*-Architecture are also contained in this level. The End-Users consist of the primary and tertiary sectors of the economy, including agriculture, transport, industry, households, services, etc. The primary coupling in the customer plant level happens between the electricity and the heating and cooling equipment available in the End-User premises.

Electricity Storage in the era of integrated energy systems takes on a new meaning. The traditional understanding of the storage that it injects electricity back at the charging point does not apply to all cases. Therefore, the storage is categorised as follows.

3.2.1.1 Storage-Link Categorisation

For transparent monitoring of storage and its appropriate consideration in algorithms of various applications of management systems, the Storage-Link is classified as follows:

- **Cat. A**: The stored energy is injected at the charging point of the grid, such as pumped hydroelectric storage, stationary batteries, etc.
- **Cat. B**: The stored energy is not injected back at the charging point on the grid, such as Power-to-Gas (P2G), batteries of e-cars, etc.
- **Cat. C**: The stored energy reduces the electricity consumption at the charging point in the near future, such as cooling and heating systems (consuming devices with energy storage potential) (Table 3.1).

3.2.1.2 Embedding P2G in the Technical/operation Level of Architecture

Figure 3.3 shows the cross-coupling of the electricity and gas vectors in the technical/functional level of the *LINK*-Architecture (see Sect. 2.5.3.1) for a typical European approach. TSOs usually operate the Extra-and Very High Voltage levels in the European type of power industry, while the DSOs operate the HV-, MV- and LV levels; e.g., Poland, Austria, etc. But some countries in Europe have a different

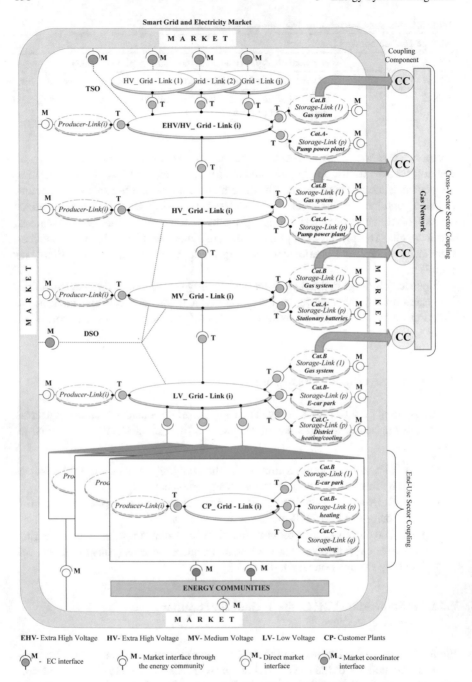

Fig. 3.3 Cross-coupling of the electricity and gas vectors in the technical/functional level of the *LINK*-architecture for a typical European approach

power industry organization that is more similar to the North American one. DSOs operate only the MV and LV levels of the grid, e.g., Italy, Fig. 3.4. In all cases, Sector Coupling takes place via the Coupling Components (CC).

LINK-Architecture spreads over the whole power grid, including high-, medium- and low voltage levels, customer plants, and the market. Storage in each voltage

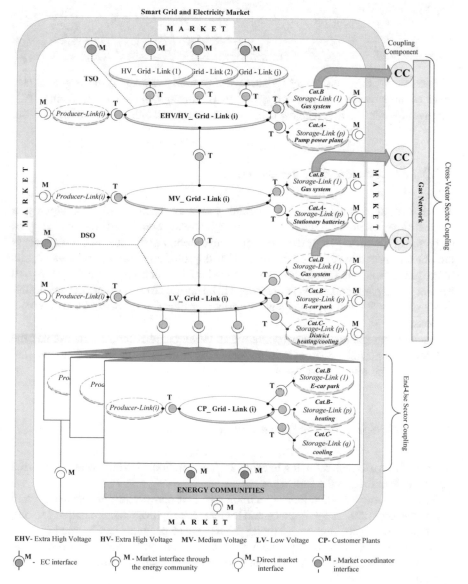

Fig. 3.4 Cross-coupling of the electricity and gas vectors in the technical/functional level of the *LINK*-architecture for a typical North American approach

level splits in Cat. A, B, and C according to the categorisation given above. The cross-coupling between the electricity and gas vector may be realised in HV-, MV- and LV levels through the storage of category B.

The main principle of the *LINK*-Solution is optimising the whole Smart Grids by coordinating and adapting the locally optimised Links.

> By the *LINK*-**Solution**, the optimisation of the electricity system and other systems is realised by coordinating and adapting the locally optimised systems.

Sector Coupling may affect (positively, if properly managed) the way ancillary services are provided and traded or valorised through market mechanisms. This positive effect is illustrated below using the process of price-driven demand response.

3.2.2 Sector Coupling Response

The activation of residential, commercial, and small business sectors, which join the real-time pricing demand response through already concluded contracts, may be triggered at any time. Their degree of participation in the demand response process may be different depending on the time of the day, duration interval, price value, etc. In the case of a surplus of electricity in the market, the electricity price decreases. All market participants and market operators will be notified to allow them the possibility to act on time. Figure 3.5 shows the information flow during price-driven demand response.

Each of the stakeholders participating in the demand response process has the possibility to perform Sector Coupling under given conditions. In the following are discussed some relevant use cases.

3.2.3 Price-Driven P2X Use Cases

A generalized use case referring to more abstract relationships and some specialised cases are described in the following.

The term power-to-X refers to technologies that convert surplus electricity into alternative products. The "X" may stand for fuel (e.g., hydrogen, methane, and so on), ammonia, chemicals, and heat, among others. These products can be used directly or converted back to electricity, making this option a flexible way to link power and fuel networks and effectively integrate intermittent renewable resources into energy systems and services [10].

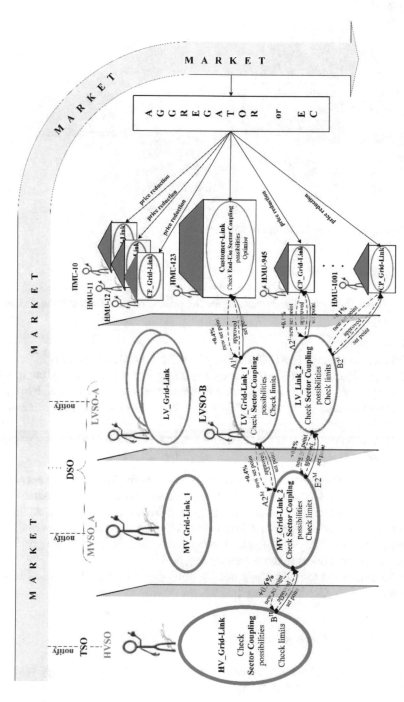

Fig. 3.5   The information flow during the process price driven demand response considering sector coupling

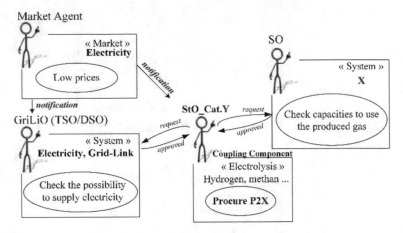

Fig. 3.6 Generalised use case: Price-driven P2X response

3.2.3.1   Generalised Use Case

Figure 3.6 shows the generalised use case for the price-driven P2X response. It has four main players:

- The market agent;
- The Grid-Link operator (GriLiO) that may be a TSO, DSO, or customer;
- The Storage Operator (StO): all Sector Coupling systems use coupling components such as electrolysis to produce hydrogen, methane, and so on. It may contribute as storage of category Y (where Y can be one of the categories A, B, or C defined above); and
- The System Operator (SO) of the respective sector X (where X can be gas system, chemical industry, etc.).

By electricity surplus, prices in the market decrease, and the Market Agent notifies the StO_Cat.Y and the GriLiOs. StO_Cat.Y requests to increase the gas production (hydrogen, methane, etc.) to the SO of X. The latter, having checked the capacities in its system, approves the action. Additionally, StO_Cat.Y checks the possibility of increasing the electricity consumption at the connection point with the grid by sending the relevant request to the corresponding GriLiO.

The latter authorises the increase in consumption after checking the limits in its grid. Finally, the StO_Cat.Y procures the P2X.

3.2.3.2   Traditional Storage Behaviour

Storage facilities that behave traditionally are part of the electricity energy vector and inject the electricity at the charging point of the grid, Cat. A.

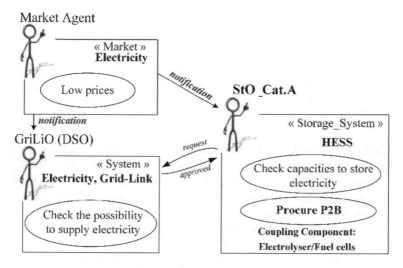

Fig. 3.7 Use case: Price-driven P2HESS response

StO_Cat.A—Power to Hydrogen Energy Storage System

Hydrogen Energy Storage System (HESS) combines an electrolysis unit, which produces hydrogen by consuming cheap electricity, and fuel cells producing elec-tricity from hydrogen. It typically has one operator. Therefore, the "Electrolyses" and "System" blocks of generalised use cases are merged in the unique block "Storage_System,". HESS is a category A storage because the electricity used to produce hydrogen is injected at the same point of the grid. Figure 3.7 shows the price-driven Power to HESS (P2HESS) response.

StO_Cat.A—Power to Batteries

Nowadays, large batteries are experimentally connected in some distribution networks to increase network flexibility. They consist of many electrochemical cells that store chemical energy to electricity and vice versa transform the chemical energy to electricity (electricity-energy → chemical-energy → electricity-energy). Coupling components are anodes and cathodes. The energy transformations happen in one device, the "Storage_System" Battery. The latter is a category A storage because the electricity used to produce hydrogen is injected at the same point of the grid. Figure 3.8 shows the price-driven Power to Battery, P2B, response.

StO_Cat.A—Power to Water.

Figure 3.9 shows the use case: Price-driven Power to Water, P2W, response. P2W is classified in the Cat.A. In this case, the StO_Cat.A operates the Pumped Hydroelectric Storage (PHES). After being informed by the market agent about the low electricity prices, PHES sends a request to the GriLiO (TSO) to increase the electricity consumption. After verifying the effectiveness of this action in its Link, the TSO approves the request, and PHES procures the P2W subsequently.

3.2.3.3   Cross-Vector Sector Coupling

Cross-Vector Sector Coupling takes place between the various vectors of the secondary economic sector. It takes place between the networks of the electric utilities and other utilities such as gas, fuel, heating, and so on. In this case, the stored energy is not injected back at the charging point on the grid, Cat.B.

StO_Cat.B—Power to Gas

Figure 3.10. shows the price-driven Power to Gas, P2G, response. It is classified in the Cat. B of the storage. The coupling component is the electrolyser. The operator of the storage Cat.B, who is responsible for H_2 production, receives from the Market Agent a notification for low electricity price. He simultaneously sends a request to the Gas System Operator, GSO, to increase, e.g., the H_2 injection and the GriLiO, to increase the electricity consumption in the connection point of the electrolyser with the Link-Grid. After having checked the availabilities in the Gas System, GSO approves the request of StO_Cat.B. The GriLiO approves the request after having verified the effectiveness of this action in the relevant Link-Grid. Consequently, the StO_Cat.B procures the P2G.

3.2.3.4   End-Use Sector Coupling

End-Use Sector Coupling happens in customer plant levels. The End-Users consist of the primary and tertiary sectors of the economy, including agriculture, transport, industry, households, services, etc. The stored energy reduces the electricity consumption at the charging point in the near future, Cat.C.

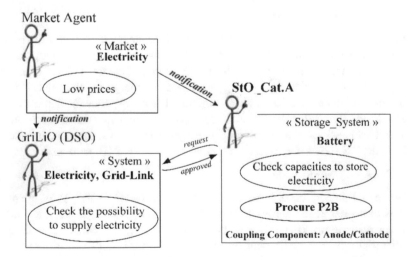

Fig. 3.8   Use case: Price-driven P2B response

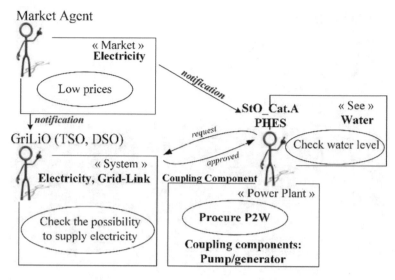

Fig. 3.9 Use case: Price-driven P2W response

From the grid perspective, customer plants in *LINK*-Solution act as black boxes (see Sect. 2.3.4.3). Within the customer plants or end-users facilities, there may be the potential to implement different types of Storage-Links.

StO_Cat.C—Power to Households

Combined Heat and Power, CHP, typically developed for residential heating, can provide ancillary services for the power grid through the Prosumer/Customer Plants.

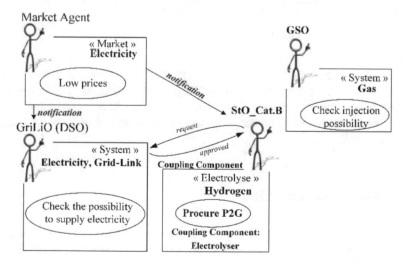

Fig. 3.10 Use case: Price-driven P2G response

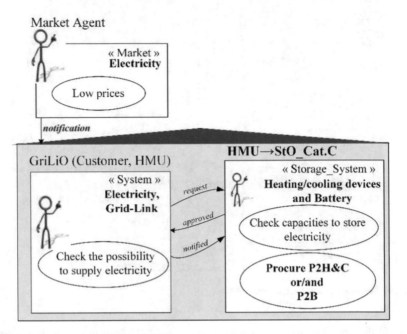

Fig. 3.11 Price-driven power to households response

The task of the secondary control, which the CP_Grid-Link establishes, is to find the optimum operating point of all available devices within the customer plant and the actual grid circumstances. Figure 3.11 shows a typical use case of the End-Use Sector Coupling: Price-driven Power to Heat and Coupling (P2H&C) response. The House Management Unit fulfils the task of the GriLiO by taking over the coordination and optimisation of the processes within the customer plant.

StO_Cat.C—Power to Chemicals

Power-to-Chemicals (P2Ch) refers to using electrical energy via water electrolysis and other downstream steps to manufacture chemical raw materials. In this way, the production of the latter may decarbonise. The products produced by electrolysis are not used for direct energy storage. However, it deals with virtual stored energy, which reduces the electricity consumption at the connection point in the near future, storage category C, Fig. 3.12.

3.3   Energy Communities

Electricity systems are facing challenges of unprecedented proportions. In the coming decade, a very substantial proportion of the electricity generation would become largely decarbonised. Beyond 2030, it is expected that significant segments of the

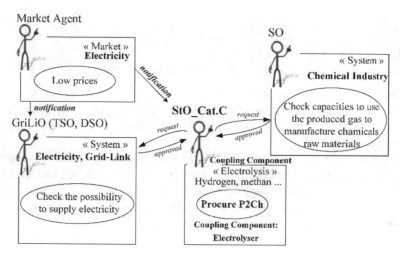

Fig. 3.12 Use case: Price-driven P2Ch response

heat and transport sectors will be decarbonised. Furthermore, in response to growing concerns associated with the security of energy supply, there is a growing interest in enhancing supply reliance based on local energy resources and smart decentralised control.

In this context, the coordination of distributed energy resources providing flexibility-related services could facilitate cost-effectiveness and secure system evolution. DER integration at several voltage levels creates replicative managing schemes to meet the demand and the technical constraints of the grid and enables the development of energy communities. It also leads to the extension of the *LINK* holistic architecture of the system by clarifying the relation between different entities and voltage levels and reassessing the data exchange between different sectors.

Worldwide, there are many "energy community" related concepts like in EU the "Local Energy Communities" [14], the "Renewable Energy Community" [5], the "Citizen Energy Community" [6]. In the USA are developed "Advanced Energy Communities" [4], in Japan "Smart Communities" [13], and in Australia "Community Energy" [1], etc. (see Appendix A.3.2) All these concepts differ only in their formulation, they have a common basis and goal. Their scope includes prosumers and DERs, without defining the voltage level (medium or low voltage level or both). In each of them, participation in the energy market is emphasised without discriminating or affecting the market participants and competition. Additionally, in each of them, the focus is set on the benefits of the whole community rather than individuals.

By definition, **Energy Communities** are no-profit cooperatives or organizations that include prosumers and DERs. They should enable participation in the energy market without discriminating neither impairing the market and competition. They should bring benefits to the community participants and should represent their interests.

The development and the implementation of energy communities face many challenges and obstacles, such as technical, regulatory, legal, economic, and stakeholder-related issues [15]. The use of holistic architecture helps to solve these challenges and obstacles rationally.

This section firstly discusses the overall EC concept and the facilities' ownership. Secondly, ECs with and without facilities ownership are embedded into the *LINK*-Architecture. It arranges Energy Communities in the centre of decision-making processes regarding future system operation and development while simultaneously delivering significantly more competitive electricity markets.

3.3.1 Facilities Ownership of Energy Communities

Concerning the EC's ownership of the electrical facilities, two cases are distinguished: EC with and without ownership on facilities as follows:

- ECs are organised in cooperatives with ownership of the electrical facilities. Figure 3.13a shows an overview of an EC with ownership of electrical facilities. In this case, the EC acts as a vertically integrated company that owns and operates the distribution infrastructure and DERs.
- ECs are organised in associations or partnerships without ownership of electrical facilities. Figure 3.13b shows an overview of an EC without ownership of electrical facilities. In this case, ECs do not own or operate any electric facility. DSOs and DERs owners operate their facilities based on schedules approved by the corresponding EC.

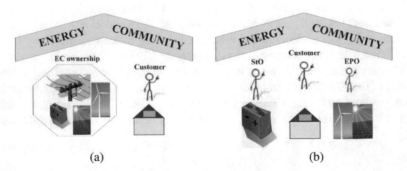

(a)   (b)

Fig. 3.13 Facilities ownership of ECs: **a** EC with ownership; **b** EC without ownership

The unified *LINK*-based holistic architecture supports the embedment of ECs in both cases, as shown in the following section.

3.3.2 Embedding the EC in the LINK-Architecture

Figure 3.14 shows the embedding of an EC with ownership on electrical facilities into the *LINK*-Architecture [12]. HV, MV, LV, and CP_Grid-Links are arranged throughout the Smart Grids. Storage and Producer-Links are connected to all Grid-Links through technical interfaces "T". The electricity market surrounds the whole

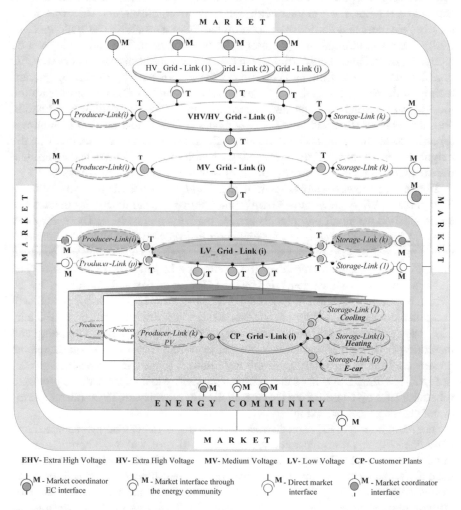

EHV- Extra High Voltage   HV- Extra High Voltage   MV- Medium Voltage   LV- Low Voltage   CP- Customer Plants

M - Market coordinator EC interface   M - Market interface through the energy community   M - Direct market interface   M - Market coordinator interface

Fig. 3.14 EC with ownership of electrical facilities embedded in the *LINK*-architecture

technical/functional architecture. It enables all stakeholders to participate in a non-discriminatory way through market interfaces "M". All grid operators, both TSO and DSO, indicated by orange interfaces in the figure, coordinate the market to guarantee a reliable and secure functioning of the own Grid-Links.

The EC organisation, marked green, surrounds all electrical facilities of its participants. EC owns and operates the LV_Grid-Link, Producer- and Storage-Links.

Prosumers participating in the EC are characterised by a dashed area. They act as black boxes to the EC, disclosing only the difference between their own production and consumption. Thus, the data privacy of each customer is guaranteed. A local electricity market (see Sect. 2.5.7.1) can be established in which the participants are all EC-cooperative members. There, they can trade electricity with each other. The EC energy surplus or deficit can be traded throughout the market, with each EC acting as a retailer. Each market actor, i.e. consumers, prosumers, DER operators' service providers that do not participate in the EC organisation, can trade its electricity directly or through an aggregator to the whole market. Thus, all market actors, members, and EC members participate in the market in a non-discriminatory way. EC operates the LV_Grid-Link and coordinates the local electricity market by guaranteeing a reliable, economical, and secure electricity supply. EC is responsible for the local long-term supplying and storage planning and the corresponding coordination with the relevant DSO and with the neighbours (EC or Grid-Links) to guarantee a reliable power supply in its area. The whole EC area acts as a black box opposite the neighbour Grid-Links. EC should exchange the required information and coordinate the operation (ancillary services provision) with neighbouring Grid-Links.

Figure 3.15 shows the embedment of an EC without ownership of electrical facilities in the *LINK*-Architecture. Like the case above, a local electricity market can be established where all participants are EC partnership members. There, they can trade electricity with each other. Special coordination with the LV_Grid-Link operator is required to keep the power supply quality such as voltage and not overload the grid. In this case, a mechanism for leasing a grid part may be necessary. The real-time schedules will be generated by the EC-local market and coordinated by the operators of the neighbouring Grid-Links. The LV_Grid-Link operator (DSO) will check the technical feasibility of the schedules and approve or reject them. EC may take over responsibilities for the local long-term supply and storage planning and the corresponding coordination with the neighbouring Grid-Links to guarantee the power supply in its area. Using Secondary Controls (on active and reactive power) of *LINK*-Solution, DSOs have the technical possibility to comply with the proposed schedules ECs by coordinating the operation with the relevant TSOs (also in terms of ancillary services provision).

The main impact of EC is on distribution systems, both in planning and operation, while the effect on transmission level is indirect. If spread above a certain amount of penetration level ECs may modify the amount and directions of power flows and the daily/seasonal load patterns. Moreover, the EC may affect (positively, if properly managed) how ancillary services are provided to TSOs and traded and valorised through market mechanisms. They can also offer flexibility-related services to the DSO, facilitating energy transition strategies associated with increasing renewable

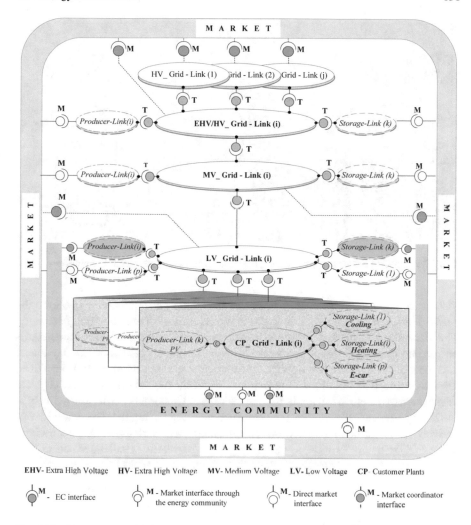

Fig. 3.15 EC without ownership of electrical facilities embedded in the *LINK*-Architecture

energy hosting capacity and promoting the deployment of electric vehicles while minimising the grid expansion investments.

3.4 Market Structure Promoting the Democratisation of the Electricity Industry

A contemporary market design is necessary to overcome the market challenges provoked by the physical limits of the transmission grid and the integration of the

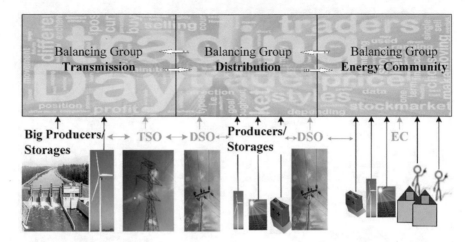

Fig. 3.16 *LINK*-markt structure promoting the democratisation of the electricity industry

distributed and renewable energy resources (see Sect. 2.5.7). Figure 3.16 shows the *LINK*-Market structure promoting the democratisation of the electricity industry. The decentralized market structure has three main parts the transmission, distribution and the energy community part. Both core market elements, the day-ahead and the real-time market are applicable in all three parts of the market. Like TSOs, the DSOs and ECs coordinate the market to prevent the malopration and physical destruction of the electrical appliances in their areas. Customers participate actively in the local market coordinated by the corresponding EC, having thus the possibility of actively participating in the price-driven demand process and of peer-to-peer trading. The *LINK*-Structure facilitates overcoming today's technical challenges because the market structure harmonises with the holistic technical model of Smart Grids.

3.5 Energy Transition to Green Economic Sectors

Climatic conditions worldwide force the comprehensive integration of renewable and distributed energy resources into the power system on a large scale, the effective use of the energy, and the existing infrastructures through effective use of Cross-Vector and End-Use Sector Coupling. The latter contributes to the effective decarbonisation of all economic sectors.

3.5.1 Scaling Up/Reproducibility

LINK-Solution can be applied to any power system worldwide. The differences between the different power supply systems are reflected in the specific *LINK*-Applications, through various parameterisation options.

3.5.2 Implementation Process

Using P2X technologies only makes sense if excess electricity from renewable energies is used from an economic and ecological perspective.

> The **large-scale integration** of renewable energy resources through the entire power grid and customer plants is a prerequisite for the comprehensive development of Sector Coupling.

The implementation of the *LINK*-Solution facilitates the large-scale integration of renewable energy resources in transmission, distribution, and customer plants level and the large-scale realisation of Cross-Vector and End-Use Sector Coupling.

Appendix

A.3.1 Economy Sectors

A nation's economy can be divided into sectors to define the proportion of a population engaged in different activities [17]. This categorisation represents a continuum of distance from the natural environment. The continuum starts with primary economic activity, which concerns itself with the utilization of raw materials from the earth, such as agriculture and mining. From there, the distance from natural resources increases as sectors become more detached from the processing of raw materials.

Primary Sector
The primary sector of the economy extracts or harvests products from the earth such as raw materials and basic foods. Activities associated with primary economic activity include agriculture (both subsistence and commercial), mining, forestry, grazing, hunting and gathering, fishing, and quarrying. The packaging and processing of raw materials are also considered to be part of this sector.

Secondary Sector
The secondary sector of the economy produces finished goods from the raw materials extracted by the primary economy. All manufacturing, processing, and construction jobs lie within this sector. Activities associated with the secondary sector include metalworking and smelting, automobile production, textile production, the chemical and engineering industries, aerospace manufacturing, energy utilities, breweries and bottlers, construction, and shipbuilding.

Tertiary Sector
The tertiary sector of the economy is also known as the service industry. This sector sells the goods produced by the secondary sector and provides commercial services to both the general population and to businesses in all five economic sectors. Activities associated with this sector include retail and wholesale sales, transportation and distribution, restaurants, clerical services, media, tourism, insurance, banking, health care, and law.

Although many economic models divide the economy into only three sectors, others divide it into four or even five. These two sectors are closely linked with the services of the tertiary sector, which is why they can also be grouped into this branch. The fourth sector of the economy, the quaternary sector, consists of intellectual activities often associated with technological innovation. It is sometimes called the knowledge economy.

A.3.2 Energy Communities

EU → Local Energy Community means an association, a cooperative, a partnership, a non-profit organisation, SME or other legal entity which is based on voluntary and open participation and is effectively controlled by local shareholders or members, the predominant aim of which is to provide local environment, economic or social community benefits for its members or the local areas where it operates rather than where it generates profits, and which is involved in activities such as distributed generation, storage, supply, provision of energy efficiency services, aggregation, e-mobility and distribution system operation, including across borders [14], the "Renewable Energy Community" [5], the "Citizen Energy Community" [6].

USA → Advanced Energy Communities are customer-focused demonstrations that integrate multiple customer resources such as Energy Efficiency, Demand Response, Customer storage, PV(or other local generation); electrification and electric vehicles in an electrically contiguous area are to achieve larger utility and societal goals such as decarbonisation, grid hardening and grid support while enabling the utility customers with advanced technologies that provide comfort, convenience, and cost benefits to the customer [4].

Japan → A Smart Community is a community where various next-generation technologies and advanced social systems are effectively integrated and utilized, including the efficient use of energy, utilization of heat and unused energy sources, improvement of local transportation systems and transformation of the everyday life of citizens [13].

Australia → Community Energy is generally defined by ARENA as being any renewable energy project initiated and/or developed by the community in order to deliver broad benefits to the local community. A community energy project may also be delivered in partnership with a private developer or independent power producer [1].

References

1. ARENA Australian Renewable Energy Agency (2018) National Community Energy Strategy. https://arena.gov.au/projects/national-community-energy-strategy-catalysing-community-ren ewables-in-australia. Accessed 05 May 2019
2. Buttler A, Spliethoff H (2018) Current status of water electrolysis for energy storage, grid balancing and sector coupling via power-to-gas and power-to-liquids: a review. Renew Sustain Energy Rev 82:2440–2454. https://doi.org/10.1016/j.rser.2017.09.003
3. Capros P et al (2019) Energy-system modelling of the EU strategy towards climate-neutrality. Energy Policy 134. https://doi.org/10.1016/j.enpol.2019.110960
4. EPRI Electric Power Research Institute (2016) Advanced energy communities: grid integration of zero net energy communities. https://eta.lbl.gov/sites/default/files/seminars/LBL_Grid_Inte gration_Presentation.pdf. Accessed 10 Nov 2020
5. EU (2018) Directive (EU) 2018/2001 of the European parliament and of the council; on the promotion of the use of energy from renewable sources, 11 December. https://eur-lex.europa.eu/ legal-content/EN/LSU/?uri=uriserv:OJ.L_.2018.328.01.0082.01.ENG Accessed 30 Mar 2020
6. EU (2019) Directive (EU) 2019/944 of the European Parliament and of the council; on common rules for the internal market for electricity and amending Directive 2012/27/EU, 5 June. https:// eur-lex.europa.eu/legal-content/EN/TXT/PDF/?uri=CELEX:32019L0944 Accessed 15 April 2021
7. European Commission (2018) A clean planet for all a European strategic long-term vision for a prosperous, modern, competitive and climate neutral economy. Brussels, pp 1–25. https://eur-lex.europa.eu/legal-content/EN/TXT/PDF/?uri=CELEX:52018DC0773&fro m=EN. Accessed 10 Jan 2019
8. EU Commission (2020) Powering a climate-neutral economy: An EU strategy for energy system integration. Brussels, pp 1–22. https://ec.europa.eu/energy/sites/ener/files/energy_sys tem_integration_strategy_pdf. Accessed 25 October 2020
9. Gadd H, Werner S (2014) Thermal energy storage systems for district heating and cooling. In: Advances in thermal energy storage systems. Woodhead Publishing Limited, Sawston, pp 467–78. 6th workshop: smart distribution management, Brussels, 13 May 2014. https://doi. org/10.1533/9781782420965.4.467
10. Hanna R, Gazis E, Edge J, Rhodes A, Gross R (2018) Unlocking the potential of energy systems integration. Energy futures lab, Imperial College London. https://imperialcollegelondon.app. box.com/s/0sil57fndc5tn9gfy6ypzp8v61qnv3mg. Accessed 06 Oct 2020
11. IEEE Task Force on Microgrid Control (2014) Trends in microgrid control. IEEE Trans Smart Grid 5:1905–1919
12. Ilo A, Prata R, Iliceto A, Strbac G (2019) Embedding of energy communities in the unified *LINK*-based holistic architecture. CIRED, Madrid, 3–6 June, pp 1–5
13. JSCA Japan Smart Community Alliance (2018) Smart Community Development. https://www. smart-japan.org/english. Accessed 07 June 2018
14. Kariņš K (2018) Report on the proposal for a directive of the European Parliament and of the Council on common rules for the internal market in electricity. Plenary

sitting A8–0044/2018, Brussels. http://www.europarl.europa.eu/sides/getDoc.do?pubRef=-//
EP//NONSGML+REPORT+A8-2018-0044+0+DOC+PDF+V0//EN. Accessed 07 Feb 2020
15. Mendes G et al. (2018) Local energy markets: opportunities, benefits, and barriers. CIRED
    workshop—Ljubljana, 7–8 June, vol 22, pp 1–4. http://www.cired.net/publications/workshop2
    018/pdfs/Submission%200272%20-%20Paper%20(ID-21042).pdf. Accessed 05 May 2019
16. O'Malley M at al. (2016) Energy systems integration: defining and describing the value
    proposition. The International Institute for Energy Systems Integration, Technical Report
    NREL/TP-5D00–66616. https://doi.org/10.2172/1257674. Accessed 08 Oct 2020
17. Rosenberg M (2020) The 5 sectors of the economy. ThoughtCo https://www.thoughtco.com/
    sectors-of-the-economy-1435795. Accessed 09 Oct 2020
18. Van Nuffel L, Dedeca JG, Smit T, Rademaekers K (2018) Sector coupling: how can it be
    enhanced in the EU to foster grid stability and decarbonise? European Parliament's Committee
    on Industry, Research and Energy. https://www.europarl.europa.eu/RegData/etudes/STUD/
    2018/626091/IPOL_STU(2018)626091_EN.pdf. Accessed 01 Oct 2020
19. Zhang X, Strbac G, Shah N, Teng F et al (2019) Whole-system assessment of the benefits of
    integrated electricity and heat system. IEEE Trans Smart Grid 10(1):1132–1145. https://doi.
    org/10.1109/TSG.2018.2871559

Chapter 4
Volt/var Chain Process*

Simplicity is the ultimate sophistication.
—Leonardo da Vinci

4.1 Introduction

Physically, the voltage is a local quantity that differs in each node of the electricity grid, as it varies along with the power transfer and the characteristics of the grid. Due to their strong coupling, the reactive power has always been used to control the voltage in transmission grids, originating the term *Volt/var process*. Reactive power is produced and consumed by diverse appliances distributed throughout the power system and operated by different stakeholders, making Volt/var control an extremely complex issue by its very nature. The overall control strategy should be designed with special care to reduce the complexity of the resulting Volt/var chain to facilitate the management.

Volt/var management is one of the most fundamental power system processes that enables the reliable and efficient operation of the electricity grids. It is used to ensure compliance to voltage limits and to keep the power factor close to one. Both the elimination of voltage limit violations and the minimisation of the reactive power flows extend the active power transfer capability of the grid. In other words, an appropriate Volt/var control increases the utilisation of the electrical infrastructure, thus lowering the need for grid reinforcement.

The massive connection of renewable and distributed generation and the electrification of other sectors gave rise to new challenges and opportunities that call for an adaption of the traditional Volt/var control schemes. Deteriorated power quality and uncontrolled interactions between the systems operated by different stakeholders are recognised as the main problems to be solved by future smart grids [66]. Meanwhile, the ability of distributed energy resources to participate in the Volt/var process

---

*Author: Daniel-Leon Schultis and Albana Ilo

© Springer Nature Switzerland AG 2022
A. Ilo and D.-L. Schultis, *A Holistic Solution for Smart Grids based on LINK–Paradigm*, Power Systems,
https://doi.org/10.1007/978-3-030-81530-1_4

[11] and the active role of customer plants [62] introduce new control opportunities. Numerous approaches are discussed in the literature that focus on certain grid parts, devices or control strategies without embedding the proposed concepts into a holistic architecture. The result is a jungle of individual solutions for individual issues that do not fit into one another to establish the optimum of the overall system.

This chapter presents the Volt/var chain process (see Sect. 2.6.2) in the framework of the *LINK*-based holistic architecture. Firstly, the corresponding physical and technical fundamentals are explained. The state-of-art is reviewed by presenting the traditional Volt/var control and the recently introduced control strategies from the holistic view of the power system. A clear problem statement is given that underlines the necessity of a holistic approach for Volt/var control. The generalised *LINK*-based Volt/var chain control is presented in great detail. Different Volt/var control setups are analysed by conducting load flow simulations in vertical power system axes. The investigated control setups are evaluated to provide a clear picture of their individual strengths and weaknesses. A general guideline and a step-by-step procedure for setting up the effective Volt/var chain control for specific vertical axes are presented.

4.2 Fundamentals of Volt/var Process

The design of an appropriate Volt/var control requires fundamental knowledge on the behaviour of the grid, the existent reactive power sources and sinks, reactive power compensation, and the available control variables and concepts.

4.2.1 Volt/var and Volt/Watt Interrelations

Flows of active and reactive power through the grid modify the node voltages in magnitude and angle, and vice versa. The two bus system shown in Fig. 4.1 is considered to clarify the interrelations between power flows and voltages. The complex impedance $\underline{Z}$ represents a series-connected grid component such as a line segment

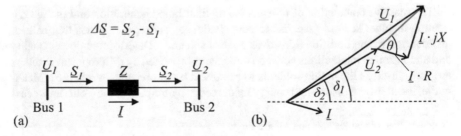

Fig. 4.1 Two bus system: **a** Circuit diagram; **b** Vector diagram

or transformer. The complex voltages and apparent power flows are designated by $\underline{U}_1$, $\underline{U}_2$ and $\underline{S}_1$, $\underline{S}_2$, respectively, and the complex current by $\underline{I}$.

The variables shown in Fig. 4.1 are expressed as in Eqs. (4.1)–(4.3).

$$\underline{Z} = Ze^{j\theta} = R + jX \tag{4.1}$$

$$\underline{U}_1 = U_1 e^{j\delta_1} \tag{4.2a}$$

$$\underline{U}_2 = U_2 e^{j\delta_2} \tag{4.2b}$$

$$\delta = \delta_2 - \delta_1 \tag{4.2c}$$

$$\underline{S}_1 = P_1 + jQ_1 \tag{4.3a}$$

$$\underline{S}_2 = P_2 + jQ_2 \tag{4.3b}$$

where Z, θ are the magnitude and angle of the complex impedance; R, X are the resistive and reactive part of the complex impedance; U_1, δ_1 and U_2, δ_2 are the magnitudes and angles of the complex voltages at buses 1 and 2, respectively; δ is the voltage angle difference between both buses; And P_1, Q_1 and P_2, Q_2 are the active and reactive power flowing into and out of the impedance, respectively. The impedance itself consumes active (ΔP) and reactive power (ΔQ) in the form of power losses, as shown in Eq. (4.4).

$$\Delta P = P_1 - P_2 = I^2 \cdot R = \frac{P_2^2 + Q_2^2}{U_2^2} \cdot R \tag{4.4a}$$

$$\Delta Q = Q_1 - Q_2 = I^2 \cdot X = \frac{P_2^2 + Q_2^2}{U_2^2} \cdot X \tag{4.4b}$$

The exact interrelations between complex power flows and voltages at both sides of the impedance are given in Eq. (4.5).

$$\underline{S}_1 = \underline{U}_1 \cdot \underline{I}^* = \frac{U_1^2}{Z} e^{j\theta} - \frac{U_1 \cdot U_2}{Z} e^{j(\theta - \delta)} \tag{4.5a}$$

$$\underline{S}_2 = \underline{U}_2 \cdot \underline{I}^* = \frac{U_1 \cdot U_2}{Z} e^{j(\theta + \delta)} - \frac{U_2^2}{Z} e^{j\theta} \tag{4.5b}$$

However, these formulas are unwieldy and do not enable an intuitive understanding of the interrelations between power flows and voltages. To estimate and

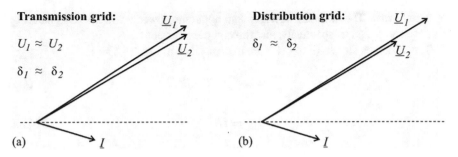

Fig. 4.2 Vector diagrams illustrating the approximations commonly made for different grids: **a** Transmission; **b** Distribution

understand the effect of power flows on the behaviour of transmission and distribution grids, simplifications are necessary. Different approximations are commonly made for transmission and distribution grids due to their distinct characteristics (R/X-ratio). The vector diagrams shown in Fig. 4.2 illustrate these approximations for both types of grids.

4.2.1.1   Transmission Grids

Partitioning Eq. (4.5b) into its real and imaginary part yields Eq. (4.6).

$$P_2 = \frac{U_1 \cdot U_2}{Z} \cos(\theta + \delta) - \frac{U_2^2}{Z} \cos(\theta) \qquad (4.6a)$$

$$Q_2 = \frac{U_1 \cdot U_2}{Z} \sin(\theta + \delta) - \frac{U_2^2}{Z} \sin(\theta) \qquad (4.6b)$$

The sensitivities of the active and reactive power flows to the voltage magnitude and angle are usually used to describe the Volt/var interrelations in transmission grids. For small deviations in voltage magnitude, the changes in active and reactive power are given by Eq. (4.7).

$$\frac{\partial P_2}{\partial U_2} = \frac{U_1}{Z} \cos(\theta + \delta) - \frac{2U_2}{Z} \cos(\theta) \qquad (4.7a)$$

$$\frac{\partial Q_2}{\partial U_2} = \frac{U_1}{Z} \sin(\theta + \delta) - \frac{2U_2}{Z} \sin(\theta) \qquad (4.7b)$$

For small deviations in voltage angle, the changes in active and reactive power are given by Eq. (4.8).

$$\frac{\partial P_2}{\partial \delta_2} = -\frac{U_1 \cdot U_2}{Z} \sin(\theta + \delta) \qquad (4.8a)$$

$$\frac{\partial Q_2}{\partial \delta_2} = \frac{U_1 \cdot U_2}{Z} \cos(\theta + \delta) \tag{4.8b}$$

Transmission grids are characterised by very low R/X-ratios, thus $\theta \approx 90°$. Regarding small differences in voltage magnitude ($U_1 \approx U_2$) and angle ($\delta_1 \approx \delta_2$) as shown in Fig. 4.2a allows approximating the voltage magnitude sensitivity of power flows, as shown in Eq. (4.9).

$$\frac{\partial P_2}{\partial U_2} \approx 0 \tag{4.9a}$$

$$\frac{\partial Q_2}{\partial U_2} \approx -\frac{U_2}{Z} \tag{4.9b}$$

Meanwhile, the voltage angle sensitivity of power flows is approximated in Eq. (4.10).

$$\frac{\partial P_2}{\partial \delta_2} \approx -\frac{U_1 \cdot U_2}{Z} \tag{4.10a}$$

$$\frac{\partial Q_2}{\partial \delta_2} \approx 0 \tag{4.10b}$$

Equations (4.9) and (4.10) indicate a strong relationship between active power flows and voltage angles; And between reactive power flows and voltage magnitudes. Hence, active and reactive power can be studied separately for many problems within transmission grids. Furthermore, Eq. (4.9b) shows that reactive power cannot be transmitted over long distances, as this would require large voltage gradients [36].

4.2.1.2 Distribution Grids

Distribution grids have much greater R/X-ratios than transmission grids. Thus, the approximation made for the impedance angle of transmission grids is no longer valid. As a result, the sensitivity considerations described above are not relevant to distribution grids. Therefore, the voltage drop equation, i.e. Eq. (4.11), is used at the distribution level.

$$U_1 \cdot \cos(\delta) - U_2 = \frac{RP_2 + XQ_2}{U_2} \tag{4.11}$$

Assuming small differences in voltage angles ($\delta_1 \approx \delta_2$) as shown in Fig. 4.2b, which is justified in distribution grids [54], yields the approximate voltage drop equation shown in Eq. (4.12).

$$\Delta U = U_1 - U_2 \approx \frac{RP_2 + XQ_2}{U_2} \qquad (4.12)$$

This formula helps to understand the Volt/var and Volt/Watt behaviour of radial distribution grids, as it allows to estimate the voltage drops along the feeders caused by the active and reactive power part of the load.

4.2.2 Reactive Power Sources and Sinks

All elements of the electric power system such as producers, storages, grid components, and consuming devices have reactive power capabilities. They produce or consume reactive power, thus acting as var-sources and -sinks, respectively.

- Producers and storages are mainly installed to contribute active power. However, they often include synchronous machines or power electronic converters whose reactive power contributions are controllable within the corresponding capability limits.
- The grid is the binding link between producers, storages and consuming devices. Its transformers and lines inherently contribute reactive power when they are energised and loaded. Additional Reactive Power Devices (RPD) are dispersed throughout the power system to increase the power factor and maintain acceptable voltages.
- Consuming devices convert electrical power into services for users. Their reactive power contributions are determined by the user and the thermostatic controls that switch on and off the consuming devices.

4.2.2.1 Producers

Producers are electricity production facilities that convert primary energy into electrical energy. Traditionally, the demand was mainly supplied by the large thermal, nuclear and hydropower plants commonly referred to as conventional power plants. In recent years, the number of Photovoltaic (PV) and Wind Turbine (WT) systems rapidly increased, constituting a significant share of the European electricity production nowadays [31]. The remaining producers, such as biofuel, solar- and geothermal, tidal power plants, etc., only slightly contribute to the overall electricity production. They are all summarised in this section under the term "others".

Conventional power plants
Conventional power plants utilise the thermal energy derived from fossil fuels and nuclear fission, and the kinetic energy of water. They use prime movers and Synchronous Machines (SM) to convert the primary energy first into mechanical and finally into electrical energy. Two main SM types are distinguished: Salient pole and cylindrical rotor. Salient pole SMs are commonly used in hydropower plants,

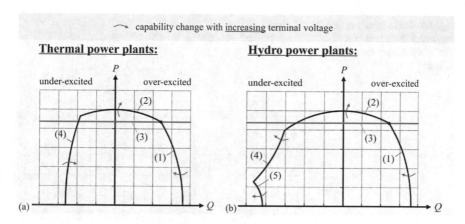

Fig. 4.3 Basic steady-state P/Q capability chart of different SM types for nominal terminal voltage:
a Cylindrical rotor; **b** Salient pole

while the cylindrical rotor ones are mainly used in thermal power plants [44]. Both convert the mechanical power at the prime mover side into active power at the grid side. The corresponding excitation systems allow regulating the machines' reactive power contributions within their P/Q capability limits. Figure 4.3 shows the basic steady-state P/Q capability chart of the cylindrical rotor and salient pole SM for nominal terminal voltage [8, 72].

In the over-excited area, the same restrictions limit the capabilities of both machine types: The maximal field current (1) and the maximal armature current (2). As the terminal voltage increases, the maximal field current narrows the capability chart, while the maximal armature current widens it [72]. Figure 4.3a shows that the under-excited region of the cylindrical rotor machine is limited by the stator end-core heating (4). The higher the terminal voltage, the more reactive power can be absorbed without overheating the stator end-core [3]. Regarding the salient pole machine, the under-excited area is restricted by the practical stability limit (4 and the minimal field current (5,When the terminal voltage increases, both limits get less restrictive [72]. In addition to the generator-related limits, the prime mover restricts the maximal active power injection independent of the terminal voltage (3).

PV systems

Photovoltaic systems basically consist of PV modules, filters and inverters. The modules convert the solar power into electrical DC power that is further converted into AC active power by the associated inverter. From the Volt/var process-related point of view, the inverter is the key component of the PV system, as it may contribute reactive power to support the grid operation. Numerous topologies of self-commutated inverters are used for the grid connection of PV modules [67]. Depending on their topology and the used modulation strategy, most inverters are capable of providing reactive power [47]. Voltage Source Converters (VSC) are the dominant inverter topology used for grid applications [17]. Within their P/Q capability limits, their

Fig. 4.4 Basic steady-state
P/Q capability chart of a
VSC for different terminal
voltages

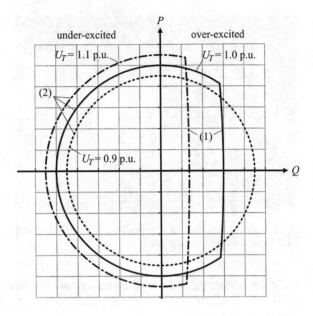

reactive power contributions can be controlled continuously and independently of
the prevalent active power conversion. The basic steady-state *P/Q* capability chart
of a VSC is shown in Fig. 4.4 for terminal voltages of 0.9, 1.0, and 1.1 p.u. [2, 5].

The *P/Q* capability of the VSC is limited by the maximum DC voltage (1) and the
maximum current through the electronic switches (2). The maximum DC voltage
significantly restricts the over-excited capability region of the inverter for high
terminal voltages. Conversely, the maximum current through the electronic switches
allows for increased power exchanges when the terminal voltage is high. The PV
module rating imposes additional restrictions on the active power capability of the
producer unit.

WT systems

WT systems are generally divided into four distinct categories based on their machine
and converter setup: Fixed-speed wind generator; Limited variable-speed wind generator; Doubly-Fed Induction Generator (DFIG); and Full-Converter Wind Generator
(FCWG) [54, 71]. Fixed- and limited variable-speed wind generators are directly
connected induction generators that consume reactive power depending on wind
speed and local voltage; Usually, fixed capacitors are added to supply the no-load
reactive power demand at nominal voltage. The converter based WT systems provide
more flexibilities. The rotors of the DFIGs are fed by back-to-back systems,[1] allowing
to independently control the active and reactive power contribution of the WT
systems. Figure 4.5 shows the steady-state *P/Q* capability chart of the DFIG for
terminal voltages of 0.95, 1.00, and 1.05 p.u. [63].

---

[1] Machine-side VSC, DC link capacitor, and grid-side VSC.

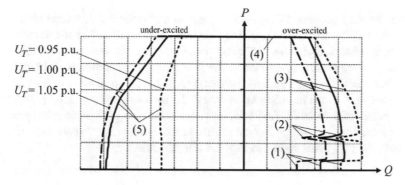

Fig. 4.5 Basic steady-state P/Q capability chart of a DFIG for different terminal voltages

The P/Q capabilities are restricted by the rated rotor voltage limit (1), rotor current de-rating at zero slip (2), rated rotor current limit (3), mechanical power limit of the wind turbine (4), and rated stator current limit (5) [26, 68]. The rotor-related limits (1)–(3) tighten with an increasing terminal voltage while the rated stator current limit widens. The FCWG may include an induction or synchronous generator connected to the grid through a back-to-back system. In this case, the WT is fully decoupled from the grid, and the resulting P/Q capability is determined by the grid side VSC, Fig. 4.4. The mechanical power limits of the WT further restrict the active power capability of the producer unit. Today, most fixed- and limited variable-speed wind generators have been replaced by DFIGs and FCWGs [71].

Others
The other producers, such as biofuel, solar- and geothermal, tidal power plants, etc., are mostly connected to the grid via SMs or VSCs [54]. These producers have similar reactive power capabilities as conventional power plants and PV systems. If induction generators are used for the power conversion, the corresponding reactive power contributions cannot be controlled as those of fixed- and limited variable-speed wind generators.

4.2.2.2 Storages

Electrical storages convert electric energy from the power system into a form that can be stored, such as electrostatic, magnetic, kinetic, potential, (electro-) chemical, and thermal energy. From the perspective of the power system, they are categorised into storages that: (A) do inject the energy back at the charging point; (B) do not inject the energy back at the charging point; And (C) reduce the energy consumption at the charging point in the near future (see Sect. 3.2.1.1). They are used for numerous applications, including portable devices, transport vehicles and stationary energy resources; Either for energy management or to improve power quality and reliability [15]. However, their capability to provide reactive power is of primary interest for the Volt/var process. Numerous storage technologies exist, including pumped hydroelectric, compressed air, battery, thermal, superconducting magnetic energy storage, flywheels, fuel cells, capacitors, etc. Pumped hydroelectric and compressed air energy storages are connected to the grid through synchronous machines, thus having similar *P/Q*-capabilities as conventional power plants. Basically, the diagrams shown in Fig. 4.3 can be mirrored to reflect the motor operating mode [72]. Meanwhile, batteries, flywheels, fuel cells and superconducting magnetic energy storage are connected via converters that determine their reactive power capabilities. If VSCs are used, the *P/Q*-chart shown in Fig. 4.4 is applicable.

4.2.2.3 Grid Components

The grid comprises lines, transformers, reactive power devices, and High Voltage Direct Current (HVDC) systems.

Lines
Overhead lines and cables are used to transmit active and reactive power from producing to consuming devices. Due to their constructive form and physical configuration, they possess resistances, inductances and capacitances per unit length. The π-equivalent circuit shown in Fig. 4.6 is commonly used to model their behaviour in load flow studies.

According to Eq. (4.13a), the reactive power consumption of a line segment depends on the current through the series inductance.

$$Q_L = Q_1 - Q_2 = X \cdot I^2 \tag{4.13a}$$

$$Q_{C,1/2} = -\omega C/2 \cdot U_{C,1/2}^2 \tag{4.13b}$$

Meanwhile, its reactive power production depends on the voltages at the shunt capacitances, Eq. (4.13b). Equation (4.13b) implies that a cable at the Medium Voltage (MV) level produces much more reactive power than one at the Low Voltage (LV) level. Table 4.1 shows benchmark parameters of European overhead lines and underground cables calculated from the data provided in [12].

Fig. 4.6 Basic π-equivalent circuit of overhead lines and cables

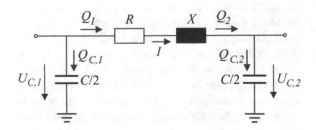

Table 4.1 Benchmark parameters of European overhead lines and underground cables

Level	Installation	Nominal voltage (kV)	R' (Ω/km)	X' (Ω/km)	C' (nF/km)	R'/X' (−)
HV	Overhead	380	0.033	0.312	11.48	0.106
	Overhead	220	0.065	0.398	9.08	0.163
	Overhead	110	0.082	0.345	10.65	0.238
MV	Overhead	20	0.510	0.366	10.10	1.393
	Underground	20	0.501	0.716	151.17	0.670
LV	Overhead	0.4	0.492	0.285	−	1.726
	Underground	0.4	0.265	0.082	−	3.232

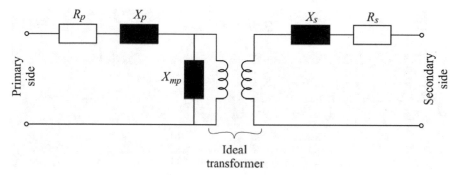

Fig. 4.7 Basic equivalent circuit of a two-winding transformer

Overhead lines generally have much lower capacitances than cables. The R/X-ratio of the wires increases from the High Voltage (HV) to the LV level. The capacitances are often neglected at the LV level as they produce only insignificant amounts of reactive power due to the low operating voltage (see Eq. (4.13b)).

Transformers

Transformers are used in the electric power system to interconnect different grid parts and to connect producers and storages with the grid. Figure 4.7 shows the basic equivalent circuit of a two-winding transformer. It includes the primary and secondary winding resistances (R_p and R_s) and leakage reactances (X_p and X_s), the magnetising reactance referred to the primary side (X_{mp}) and the ideal transformer.

Transformers always consume reactive power depending on their loading: In no or low load conditions, the effect of the shunt magnetising reactance predominates, while in high load conditions, the impact of the series leakage reactance dominates [36].

Reactive power devices

RPDs are used for many purposes, including voltage control, load compensation, power flow control, damping of oscillations, and for improving voltage, transient and dynamic stability. They are series- or shunt-connected. Series-connected RPDs are mainly[2] used for power flow control and oscillation damping. Meanwhile, the shunt-connected ones are primarily used for voltage control and load compensation. However, both types affect the reactive power flow and consequently the grid voltages [29].

$$Q_{RPD} = Q_1 - Q_2 \tag{4.14}$$

Figure 4.8 and Eqs. (4.14)–(4.18) illustrate the physical principles of reactive power contribution.

Figure 4.8a, b illustrate the principles of the series reactive power contribution. In Fig. 4.8a, a series voltage U_{RPD} is injected in quadrature to the line current I, provoking the reactive power contribution according to Eq. (4.15).

$$Q_{RPD} = I \cdot U_{RPD} \tag{4.15}$$

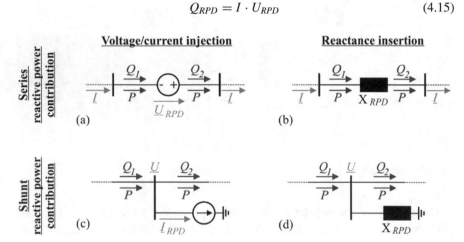

Fig. 4.8 Physical principles of reactive power contribution: **a** Series voltage injection; **b** Series reactance insertion; **c** Shunt current injection; **d** Shunt reactance insertion

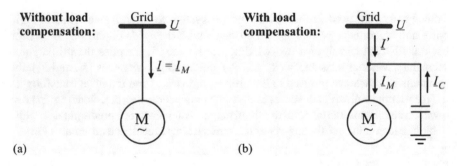

Fig. 4.9 Induction motor: **a** Without load compensation; **b** With load compensation at the device level

---

[2] Series reactive power contribution is sometimes used in distribution level to secure the supply of factories against voltage dips [79].

In Fig. 4.8b, a series reactance X_{RPD} is inserted that contributes reactive power according to Eq. (4.16).

$$Q_{RPD} = I^2 \cdot X_{RPD} \qquad (4.16)$$

This formula is compliant with Eq. (4.4b), which describes the reactive power losses of a series connected impedance. The shunt reactive power contribution principles are illustrated in Fig. 4.8c, d. In Fig. 4.8c, a shunt current I_{RPD} is injected in quadrature to the node voltage U, resulting in the reactive power contribution according to Eq. (4.17).

$$Q_{RPD} = U \cdot I_{RPD} \qquad (4.17)$$

The inserted shunt reactance X_{RPD} shown in Fig. 4.8d contributes reactive power depending on the node voltage, Eq. (4.18).

$$Q_{RPD} = U^2 / X_{RPD} \qquad (4.18)$$

Numerous RPDs have been developed that take advantage of the physical principles outlined above.

Table 4.2 groups the most widely used ones according to the physical principle they rely on; Their reactive power capability and the corresponding controllability are also given. A detailed description of the listed RPDs can be found in [23] and [29].

Fixed shunt capacitors and reactors are permanently connected to the grid, while the mechanically switched ones (MSC and MSR, respectively) are connected via circuit breakers. MSCs are often split into smaller capacitor units to improve the smoothness of their control. As the constructional parameters determine their maximum capacitance, their reactive power capability strongly depends on the local voltage (see Eq. (4.18)). For low voltages, i.e. when the reactive power support is most needed, their reactive power capability is very low. Static Var Compensators (SVC) employ Thyristor-Controlled Reactors (TCR), Thyristor-Switched Reactors (TSR) and Thyristor-Switched Capacitors (TSC) to continuously adjust their reactance. Therefore, analogously to the MSC, the SVC has deficient reactive power capabilities for low voltages.

Static Synchronous Compensators (STATCOM) are shunt-connected self-commutated converters whose output current can be continuously controlled independently of the terminal voltage. According to Eq. (4.17), their reactive power capabilities decrease linearly with the terminal voltage, thus offering better voltage support for low voltages as the SVCs. Synchronous Condensers (SyC) are synchronous machines running without prime mover or mechanical load. They inject reactive currents in quadrature to the node voltage.

The fixed series capacitor produces reactive power depending on the square of the current flowing through it, Eq. (4.16). It compensates for the line series inductance, reducing the effective reactive power loss (see Eq. (4.4b)). In this regard, the fixed series capacitor acts self-regulating [36]. The Thyristor-switched series capacitor (TSSC) and reactor (TSSR) consist of a series capacitor and reactor, respectively, shunted by a TSR. They allow for stepwise reactance control. Smoothly variable reactance is provided by both Thyristor-Controlled Series Capacitor (TCSC) and Reactor (TCSR), as they use thyristor-controlled reactors instead of the thyristor-switched ones.

The Static Synchronous Series Compensator (SSSC) resembles the STATCOM, except that it injects a voltage in series with the line, Eq. (4.15). Its reactive power capability linearly depends on the line current.

Additional active power sources or storages may be connected at the DC side of STATCOMs and SSSCs, creating the ability to inject and absorb active power [29]. This topic is not treated in this chapter because this ability is not related to the Volt/var control process.

HVDC systems

HVDC systems are used for long-distance active power transmission and to interconnect asynchronous AC systems. They basically comprise two transformers, two power electronic converters, and DC transmission lines [1]. The absence of capacitive charging effects at the DC lines allows the low-loss submarine, underground

Table 4.2 Overview of reactive power devices

Principle	RPD	Acronym	Capability	Controllability
Shunt reactance insertion	Fixed shunt capacitor	–	Capacitive	None
	Fixed shunt reactor	–	Inductive	None
	Mechanically switched capacitor	MSC	Capacitive	Discrete
	Mechanically switched reactor	MSR	Inductive	Discrete
	Static var compensator	SVC	Both	Continuous
Shunt current injection	Static synchronous compensator	STATCOM	Both	Continuous
	Synchronous condenser	SyC	Both	Continuous
Series reactance insertion	Fixed series capacitor	–	Capacitive	None
	Thyristor-switched series capacitor	TSSC	Capacitive	Discrete
	Thyristor-switched series reactor	TSSR	Inductive	Discrete
	Thyristor-controlled series capacitor	TCSC	Capacitive	Continuous
	Thyristor-controlled series reactor	TCSR	Inductive	Continuous
Series voltage injection	Static synchronous series compensator	SSSC	Both	Continuous

and overhead power transmission. Two main categories of HVDC converters are in use: Line-Commutated Converters (LCC) and VSCs. Both convert AC active power into DC one, and vice versa. However, their reactive power capabilities substantially differ from each other. The LCCs inherently consume reactive power of about 50–60% of the prevalent active power conversion. Therefore, supplementary reactive power sources are usually installed nearby the converter stations to compensate for the reactive power demand at nominal conditions. Meanwhile, the VSCs provide more flexibilities as they allow independent control of their reactive power contributions (see Sect. 4.2.2.1). The DC power lines add active power limits to the P/Q capability chart shown in Fig. 4.4. Today, LCCs still dominate the HVDC market. However, a shift towards VSC technology is already noticeable and expected for the future [1].

4.2.2.4 Consuming Devices

Consuming devices convert electric power into services for the users. They consume active power and contribute reactive power when they are switched on. Traditional consuming devices such as directly connected induction motors behave inductively. In contrast, many modern ones, such as switch-mode power supply loads with passive power factor correction, adjustable speed drives, compact fluorescent lamps, and Light Emitting Diodes (LED), show capacitive behaviour [58]. When switched on, their power consumption depends on the supplying voltage (Preiss and [48]. In load flow studies, this voltage dependency is commonly modelled using the exponential or the polynomial (also called "ZIP") load model [4, 49]. The polynomial load model relies on Eq. (4.19).

$$P^{Dev}/P^{Dev}_{nom} = C^{Z,P} \cdot (U/U_{nom})^2 + C^{I,P} \cdot (U/U_{nom}) + C^{P,P} \qquad (4.19a)$$

$$Q^{Dev}/Q^{Dev}_{nom} = C^{Z,Q} \cdot (U/U_{nom})^2 + C^{I,Q} \cdot (U/U_{nom}) + C^{P,Q} \qquad (4.19b)$$

$$C^{Z,P} + C^{I,P} + C^{P,P} = C^{Z,Q} + C^{I,Q} + C^{P,Q} = 1 \qquad (4.19c)$$

where $C^{Z,P}$, $C^{I,P}$, $C^{P,P}$ and $C^{Z,Q}$, $C^{I,Q}$, $C^{P,Q}$ are the active and reactive power-related ZIP coefficients; P^{Dev}, Q^{Dev} and P^{Dev}_{nom}, Q^{Dev}_{nom} are the active and reactive power consumptions for the actual and nominal voltage, respectively; And U, U_{nom} are the actual and nominal supplying voltage, respectively. However, considering the power system's fractal structure [34], the consuming devices themselves consist of several smaller elements, i.e., conductors, transformers, reactive power devices and active power appliances. Hence, they can be regarded as a grid part such as the Customer Plant (CP) or low voltage grid, but smaller in size.

4.2.3   Reactive Power Compensation

Reactive power compensation contains two aspects: load compensation and voltage support [23].

4.2.3.1   Load Compensation

Load compensation is traditionally used to locally supply the reactive power demand of industrial customer plants and consuming devices such as motors and discharging lamps [50].

> The goal of load compensation is to increase the power factor of industrial CPs and consuming devices independent of their supplying voltage.

The reactive power demand of induction motors is locally supplied by capacitors connected at the CP level or directly at the device level. In the former case, the capacitor may supply the reactive power demand of many motors. Figure 4.9a, b show an induction motor without and with load compensation at the device level, respectively. The motor is connected to a bus bar with the supplying voltage $\underline{U}$.

Without load compensation, the current drawn from the grid $\underline{I}$ equals the current consumed by the motor $\underline{I}_M$, Eq. (4.20a). When load compensation is used, the grid current $\underline{I}$ is composed of the motor $\underline{I}_M$ and the capacitor current $\underline{I}_C$, Eq. (4.20b).

$$\underline{I} = \underline{I}_M \tag{4.20a}$$

$$\underline{I}' = \underline{I}_M - \underline{I}_C \tag{4.20b}$$

The vector diagrams of both cases are shown in Fig. 4.10. When no load compen-

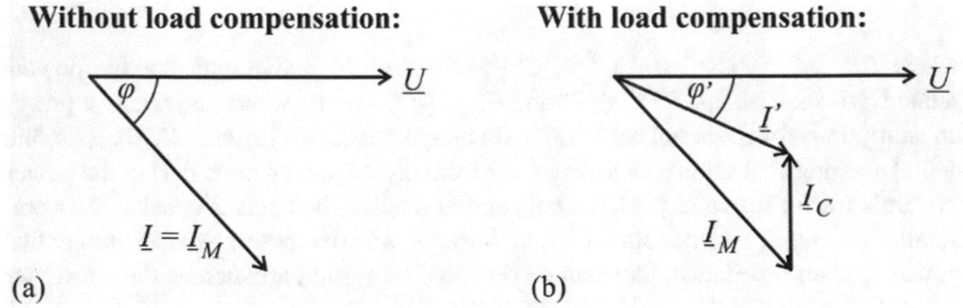

Without load compensation: **With load compensation:**

(a)   (b)

Fig. 4.10 Vector diagrams of induction motor: **a** Without load compensation; **b** With load compensation at the device level

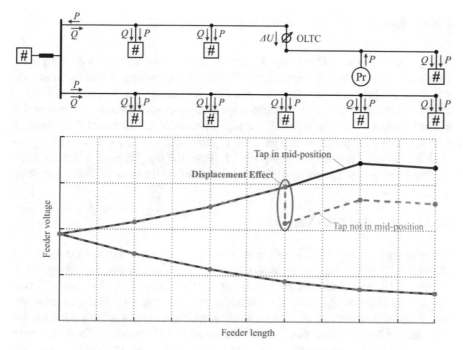

Fig. 4.11 Effect of an LVR's tap position on the voltage profiles of two distribution feeders

sation is applied, the grid current lags the voltage by a significant angle φ, leading to a poor power factor at the grid connection point, Fig. 4.10a.

Figure 4.10b shows that the reactive current injection of the capacitor partly compensates for the consumption of the motor, decreasing the angle φ and thus improving the power factor at the bus bar. Furthermore, the magnitude of the grid current is reduced.

> Load compensation reduces the reactive power flows in the grid and thus its loading, losses and voltage drop. The reduction of line and transformer loading increases the active power transfer capability of the electricity grid.

4.2.3.2 Voltage Support

Reactive power is traditionally compensated at the HV and MV levels to maintain acceptable voltages and increase the power factor throughout the grid. The grid voltage and the reactive power flows are closely coupled. Therefore, they are controlled together in the course of the Volt/var process, described comprehensively in this chapter.

4.2.4   Volt/var Control Variables

The grid voltages are affected by the active and reactive power flows and the tap positions of various transformers. Active power is the good that should be transferred and distributed and does not serve to control the voltage. As a result, only two Volt/var control variables remain, i.e. the shunt var contributions that have a considerable indirect impact on the voltage and the tap positions of transformers that control it directly. Series var contributions as those of series-connected RPDs are usually not used for Volt/var control (see Sect. 4.2.2.3). Both the tap positions of transformers and the shunt var contributions differ in their impact on the feeders' voltage profiles.

4.2.4.1   On-Load Tap Changers

Tap changers are usually implemented in every transformer, but the On-Load Tap Changers (OLTC) are typically available in power transformers, Supplying Transformers (STR) and Line Voltage Regulators (LVR). Meanwhile, Distribution Transformers (DTR) usually have tap changers that can only be changed after de-energisation. The control of the tap position varies the transformer's transmission ratio, provoking a voltage magnitude jump. Figure 4.11 illustrates the effect of an LVR's tap position on the voltage profiles of two distribution feeders. The change of the tap position shifts a part of the voltage profile mainly in parallel to the original one. This effect is denoted as the Displacement Effect [32]. The parallelism is usually inaccurate due to the voltage dependency of the consumption (see Sect. 4.2.2.4) and feeder losses (Sect. 4.2.2.3).

4.2.4.2   Shunt Var Contributions

Shunt var contributions provoke reactive power flows through the grid that modify the grid voltages (see Eq. (4.12)). Figure 4.12 illustrates the effect of the var absorption of a shunt-connected RPD on the voltage profiles of two distribution feeders. The change of the reactive power contribution has two effects on the resulting voltage profiles [32]:

- The Displacement Effect shifts the voltage at the supplying bus bar. As a consequence, the voltage profiles of both feeders are moved in parallel.
- The Push Down/Up Effect pushes the voltage profile down or up for inductive or capacitive var contributions, respectively. The maximal shift occurs at the connection point of the RPD. This effect rotates the voltage profile of the feeder to which the RPD is connected.

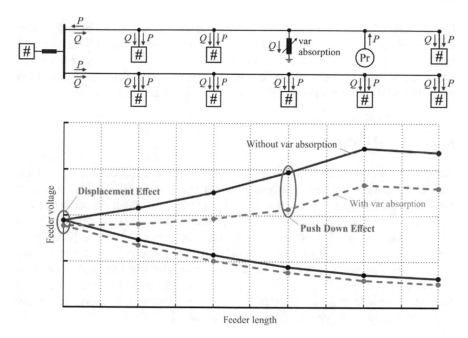

Fig. 4.12 Effects of the var absorption on the voltage profiles of two distribution feeders

4.2.5 Volt/var Control Concepts

Volt/var control is an essential process for system operators to ensure the reliable and efficient operation of the power system. This process aims to operate the existing control variables to maintain acceptable voltages optimally. The management of the var contributions of producers, storages and RPDs is challenging due to their vast number, distributed nature, discrete or continuous adjustability, and the complexity of the physical laws that determine the overall system behaviour. Different parts of power systems are owned and operated by distinct stakeholders. Since the power system is a physical entity and the control actions within a particular grid part also impact the system behaviour in other parts of the grid, the overall control strategy must be agreed between the involved stakeholders. Different concepts are used to control the available variables, including manual, local, primary, direct and secondary control (see Sect. A.2.2).

4.2.5.1 Manual Control

Manual Control (MC) is used to adjust discrete control variables such as the tap positions of transformers and the switch positions of mechanically switched capacitors. As a prerequisite, distributed measurements are collected via the Supervisory Control And Data Acquisition (SCADA) system and visualised in the control centre.

On this basis, the operator manually initiates the control action required to achieve the desired system state. For example, the Transmission System Operator (TSO) recognises a deviation from the scheduled reactive power exchange with the neighbour transmission grid and manually changes the switch position of the remotely controlled MSC or authorises a crew to execute the command on site.

4.2.5.2 Local Control

Local Control (LC) refers to control actions that are carried out locally without considering the holistic real-time behaviour of the relevant grid part. Its action path may be realised in open- or closed-loop. LC automatically adjusts the active/reactive power contributions of RPDs, storages and producers, and the tap positions of transformers based on local measurements or time schedules [28, 53, 66]. It usually maintains a power system parameter, which is locally measured or calculated based on local measurements, equal to the desired value. The fixed control settings are calculated based on offline system analysis for typical operating conditions. Local controls are simple, reliable and quickly respond to changing operating conditions without the need for a communication infrastructure [43, 52, 81].

4.2.5.3 Primary Control

Primary Control (PC) refers to control actions executed locally in a closed-loop: The input and output variables are the same. The output- or control variable is locally measured and continuously compared with the set-point received from the corresponding Secondary Control (SC). The deviation from the set-point results in a signal that influences the valves or frequency, excitation current or reactive power, transformer tap positions, etc., in a primary-controlled power plant, transformer, and so on, such that the desired power is delivered or the desired voltage is reached.

4.2.5.4 Direct Control

Direct Control (DiC) refers to control actions performed locally in an open-loop, taking into account the real-time holistic behaviour of the corresponding grid part. The secondary control calculates the control action.

4.2.5.5 Secondary Control

Secondary control refers to control variables that are calculated based on the current state of a control area. It fulfils a predefined objective function by respecting the constraints of electrical appliances (P/Q capabilities of generators, transformer rating, voltage limits, etc.). It calculates and sends the set-points to primary controls and the input variables to direct controls acting on its area.

4.2.6 Monitoring

The Volt/var process aims to maintain acceptable voltages and high power factors throughout the entire power system by managing the available reactive power resources and on-load tap changers. This management may be established by coordinated or uncoordinated control.

Operators use SCADA systems and, in many cases, also state estimators to observe the steady-state and specify set-points for the distributed control variables to maintain limit compliance within the observed area. The set-points may be calculated automatically by secondary controls or manually by the personnel of the control centres. In any case, the coordinated Volt/var control builds upon monitoring the power system (see Sect. 2.6.1).

In contrast, the uncoordinated control uses fixed control settings for the distributed control variables determined based on offline power system analysis.

4.3 Traditional Volt/var Control

Traditionally, the power system was divided into two levels: Transmission, which includes the very high and high voltage level; And distribution, which includes the medium and low voltage levels. Customer demand was mainly met by the large thermal and hydropower plants connected at the transmission level. The traditional Volt/var applications were designed to cope with the resulting unidirectional power flows from transmission via distribution grid to the consuming devices connected at the customer plant level. Energy (EMS) and Distribution Management Systems (DMS) were used by the operators mostly in advisory mode, i.e. in open-loop. Local and manual controls are used to control the reactive power and voltage in very high, high and medium voltage grids. The low voltage grids were sufficiently dimensioned based on the estimated peak demand and the expected annual demand increase to guarantee acceptable voltages in low voltage level. Figure 4.13 shows an overview of the traditional devices equipped with Volt/var control units from the holistic view of the power system. The available control variables are highlighted in blue.

The synchronous machines located in the conventional power plants and pumped hydroelectric storages connected at the transmission level provide the basic means of voltage control. Their automatic voltage regulators act as local controls that maintain scheduled voltage set-points at the generator terminals. Additional RPDs are distributed throughout the transmission grids to maintain voltages and improve the power factor. Analogously to the generators of conventional power plants, synchronous condensers and static var compensators are mostly controlled to establish voltage set-points at their terminals. Due to their high purchasing and operating costs, the synchronous condensers have been widely replaced by static var compensators. Shunt reactors are typically used at long overhead lines to prevent the violation of the upper voltage limit caused by the line capacitances during light load

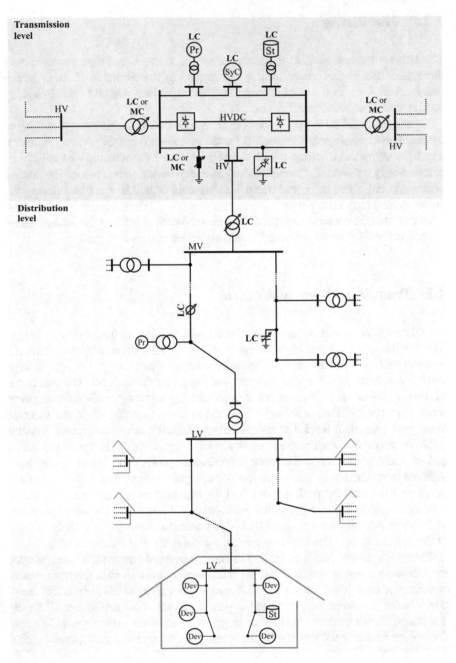

Fig. 4.13 Overview of the traditional devices equipped with Volt/var control units from the holistic view of the power system

conditions. They may be fixed and permanently connected or switched if less reactive power support is required during high load conditions. Fixed and mechanically switched capacitors are also used to supply the reactive power demand of overhead lines and subordinate distribution grids. The reactors and capacitors connected to the transmission grid are switched automatically, depending on locally measured quantities such as voltage, current and so on, or based on time schedules (**LC**). They may also be switched manually by the system operator (**MC**). The tap positions of power transformers affect the grid voltages and the reactive power exchanges between the transmission grids. They are controlled either locally to maintain one of the bus voltages or manually. HVDC systems were implemented with LCC technology, thus without the ability to adjust their reactive power contribution and to participate in Volt/var control actively. [36]

Voltage in the distribution level is controlled using the OLTC of supplying transformers, line voltage regulators and mechanically switched capacitor banks [36]. The STR and line voltage regulators are locally controlled to maintain the voltages at their secondary buses or other nodes using line drop compensation. Mechanically switched capacitors are used for power factor correction and MV feeder voltage control. They are automatically switched depending on time clocks, terminal voltage, power factor, etc. The small hydropower plants are traditionally operated with a unity power factor [76].

Storages within the customer plants (mainly heating and cooling systems) do not participate in the Volt/var control process. Traditionally, no producers are connected at the CP level.

4.4 Recently Introduced Control Strategies

In recent years, many inverter-coupled producers, including PV and WT systems, have been installed across all grid levels. Their intensive and intermittent injections pose new challenges for the power system operation that call for an adaption of the traditional Volt/var control schemes. Fluctuating voltages, increased operations of discrete control variables, and bidirectional active power flows through the vertical power system axis are the consequence. The original dimensioning of low voltage grids is insufficient to cope with the simultaneous PV injections and the high consumption of Electric Vehicle (EV) chargers, making Volt/var control necessary also at the LV level. From here, the separate consideration of medium and low voltage grids became established to identify the Volt/var control requirements of both system levels. Meanwhile, CPs are implied in the low voltage level and not considered separately.

Up to now, the main strategy used to meet the new challenges was to increase the number of control variables. The inverter-coupled producers across all grid levels are equipped with local controls to utilise their reactive power capabilities for the Volt/var process. The rapid development in power electronics gave rise to new appliances such as FACTS devices and HVDC systems in VSC technology that opened up new control possibilities at the high voltage level. Furthermore, the upgrade of DTRs with OLTCs is considered as a potential measure to increase power system controllability. Most EMS and DMS still run in open-loop, and the settings of the local controls are calculated based on offline system analysis. The recent Volt/var control is overviewed in Fig. 4.14 from the holistic view of the power system.

The STATCOMs and VSCs of HVDC systems recently connected to the HV level are locally controlled to establish voltage or reactive power set-points [24, 77]. The local controls of large WT and PV systems usually maintain constant reactive power, power factor or voltage set-points or adjust the reactive power contribution depending on the local voltage [38].

Some of the small hydropower plants connected at the MV level are upgraded with local controls that enable the voltage or power factor control mode [76]. Diverse storage technologies are installed, but they are usually not incorporated in the Volt/var process. The distributed PV and WT systems are mostly operated with fixed reactive power or power factor, or according to $\cos\varphi(P)$- or $Q(U)$-control characteristics.

Local control of the PV systems' reactive power contributions (which are mostly located at the CP level) is used to maintain voltage limit compliance at the LV level. Constant reactive power, constant power factor, $\cos\varphi(P)$- and $Q(U)$-control are the most common operating modes [9]. These modes have been further refined by including more local variables, e.g. $\cos\varphi(P,U)$ [21], $Q(U,P)$ [80], $\cos\varphi(P,R/X)$ [65], and many more [70]. Figure 4.15a shows the symbolic representation of a PV system with local Volt/var control connected to the CP grid. Alternatively to the control of PV systems, the DTR may be upgraded with an OLTC and locally controlled to maintain its secondary voltage close to a defined value. This measure may also be combined with the $Q(U)$-control of PV systems [39]. EV chargers are increasingly installed at the CP level, but they are usually not integrated into the Volt/var process.

4.4.1 Cosφ(P)-Control

A $\cos\varphi(P)$-controlled inverter absorbs reactive power when the active injection exceeds a certain value, which is commonly set to 50% of the maximal production. Usually, the power factor is reduced from 1.0 down to 0.9 inductive during periods of peak production. Alternatively, for low PV penetrations, it might be less decreased. The fundamental control characteristic is shown in Fig. 4.15b.

4.4.2 Q(U)-control

When $Q(U)$-control is applied, the PV inverter absorbs reactive power for high local voltages and injects reactive power for low ones, Fig. 4.15c. The parameters U_a, U_b, U_c, and U_d, are specified by the responsible Distribution System Operator (DSO)

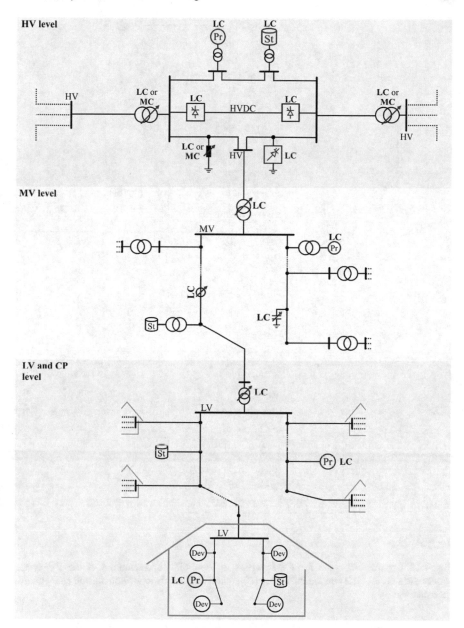

Fig. 4.14 Overview of the recent Volt/var control from the holistic view of the power system

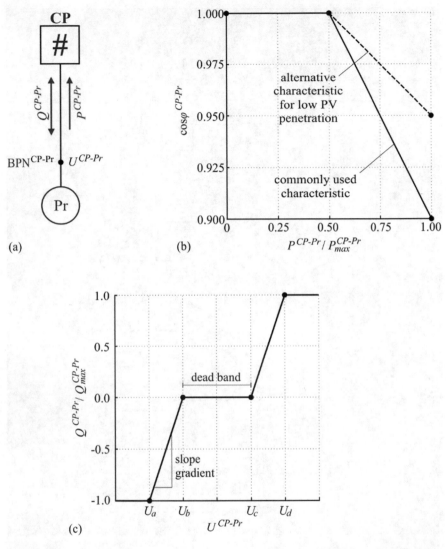

(a)

(b)

(c)

Fig. 4.15 Control strategies for PV inverters: **a** Symbolic representation of the PV system connected in CP level; **b** Fundamental $\cos\varphi(P)$-control characteristic; **c** Fundamental $Q(U)$-control characteristic

according to the prevalent grid conditions. The control parametrisation is discussed in detail in Sect. A.4.2.1.

4.4.3 OLTC in the Distribution Substation

The on-load tap changer in the distribution substation modifies the voltage at the DTR secondary bus, thus shifting the voltage profiles of all LV feeders in parallel. It is usually primary controlled to adjust its tap position when the voltage error at a certain grid node deviates from the specified dead-band during a specified time delay [16].

4.5 Problem Statement

The local controls recently implemented at LV and CP level are insufficient to meet the technical and social requirements of future Smart Grids. The OLTC in the distribution substation is slow to operate and sensitive to the number of tap operations. It cannot react appropriately to the voltage fluctuations caused by the intermittent PV injections. Temporary voltage limit violations and unnecessary tap operations are the consequence, thus jeopardising the electrical equipment and shortening the transformer's durability [30]. The use of local $\cos\varphi(P)$- and $Q(U)$-control provokes extensive uncontrolled reactive power flows throughout all grid levels [32] and makes active power curtailment necessary to ensure supply quality and reliability [69]. Implementing $\cos\varphi(P)$- and $Q(U)$-controls in PV systems intertwines the operation of the LV grids and customer plants [33]. Consequently, the reactive power is procured by non-market-based and—in the case of $Q(U)$-control—discriminatory procedures, which fundamentally contradicts [22].

Although not implemented on a large scale yet, the necessity for the closed-loop coordination of the distributed control variables is widely recognised. The coordinated automatic voltage regulation in transmission grids has already been studied in Italy, France, Belgium and Spain [19]. Still, the coordination between transmission and distribution grids has not been considered. Today, ENTSO-E proposes the use of secondary and primary controls for dynamic optimisation and voltage control at the high voltage level [24]. Furthermore, advanced DMS are emerging that allow— among many others—for the closed-loop Volt/var optimisation of distribution grids on a 24/7 basis [14]. Few distribution utilities such as 'Inland Power and Light' and 'Clatskanie PUD' [75], 'Dominion Virginia Power' [45], 'SCE' [42], 'Duke Energy' [7], and 'BC Hydro' [20], have already implemented the closed-loop Volt/var control.

However, the lack of a holistic approach makes the large scale roll-out of the new control paradigms almost impossible. Coordination schemes that do not rely on a holistic approach consider the perspective of individual stakeholders (TSOs, DSOs, and customers) and optimise individual functionalities, thus leading to suboptimal solutions from the global view. European utilities supply about 260 million CPs, of which more than 99% are connected at the LV level [27]. Coordinating the reactive power contributions of millions of CPs results in a flood of data exchange that poses severe challenges to ICT [51], cybersecurity [73] and data privacy [78].

4.6 Volt/var Chain Control

The *LINK*-Architecture allows splitting the horizontal and vertical power system axes into chains of locally optimised Links that interact with each other in a coordinated way (see Sect. 2.6.2). It accomplishes the Volt/var process by a chain control to increase the electrical infrastructure utilisation. Its effectiveness may be increased using an extended lumped grid model. Innovative control strategies at the LV and CP levels overcome the insufficiencies of the recently implemented control schemes.

4.6.1 Extension of the Lumped Grid Model

The hardware of the system components builds the skeleton of the power system that determines its fundamental physical behaviour (see Sect. 1.4). Their accurate modelling is the basis of power system analysis and the prerequisite to study the effect of control strategies on the system behaviour. However, the power system is often referred to as the biggest and most complex human-made system existing on earth owned, operated or used by different stakeholders. Its complete modelling and simulation as a single entity are impracticable due to the related modelling effort, the required computational resources and unknown system details. Therefore, power system analysis is restricted to relatively small grid parts and considers the connected neighbouring elements by lumped models. The lumped modelling of system parts is challenging as these parts usually include numerous elements with distinct characteristics and complex interdependencies.

4.6.1.1 Lumped Modelling for Power Flow Analysis

Power flow analysis is used to study the steady-state behaviour of grid parts. While the grid part under study is modelled in detail, lumped models represent the connected producers, storages, and neighbouring grid parts. Each lumped model corresponds to one of the following node-elements:

- *PV node-element*: The element's active power contribution and the voltage magnitude at its boundary node are specified for the regarded instant of time. Additional Q-limits may be defined. Typical examples include producers such as generators of conventional power plants and RPDs such as static var compensators and synchronous condensers.
- *PQ node-element*: The element's active and reactive power contributions are specified for the regarded instant of time, either independent of the voltage at the boundary node or as a function of it. Typical examples include producers such as PV and WT systems, storages, and neighbour Link-Grids.
- *Slack node-element*: The voltage magnitude and angle at its boundary node are specified for the regarded instant of time. For power flow analysis at MV, LV, and

CP (and device) level, the superordinate grid is usually selected as the sole slack node-element. Meanwhile, at the HV level, the largest generator may be defined as the slack node-element, or the slack node-element may be divided between different large generators.

The specification of the PQ node-elements that represent entire system parts is of great uncertainty. Their active and reactive power contributions vary over time and node voltage. The time dependency mainly results from the behaviour of customers and thermostatic controls that switch on and off the consuming devices, from the weather conditions that determine the injection of renewable producers, and from the schedules of storages and thermal power plants. Meanwhile, the voltage dependency inheres the power system hardware itself and may be intensified by control schemes such as $Q(U)$.

4.6.1.2 Conventional Model

Figure 4.16 shows an overview of the conventional monitoring model. The grid part under study may apply to HV, MV or LV grids, Fig. 4.16a. Neighbouring grid parts are conventionally modelled as PQ node-elements specifying their $P(U)$ and $Q(U)$ behaviour for each regarded instant of time t, Fig. 4.16b. They are commonly represented by the " $\rightarrow$ " symbol in power flow analysis tools. No voltage limits are associated with the conventional lumped models of neighbouring grid parts. Instead, narrow voltage limits are associated with the study grid part to imply safety margins for the neighbouring ones, which are represented by lumped models. For example, a maximal voltage drop of ± 0.06 p.u. is expected to appear at the LV level under worst-case conditions. The MV grid is operated within voltage limits of ± 0.04 p.u. around the nominal value to ensure limit compliance at the LV level.

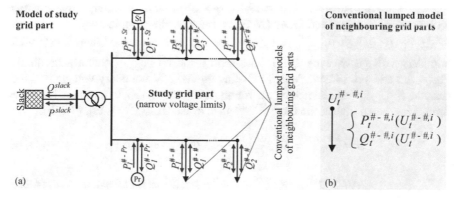

Fig. 4.16 Overview of the conventional monitoring model: **a** Detailed model of the study grid part; **b** Lumped model of neighbouring grid parts

4.6.1.3   Extended Model

The extended lumped model is explained in the following subsections by introducing the concept of Boundary Voltage Limits (BVL) and summarising all model components. Furthermore, the procedure used to calculate the extended lumped model of a Link-Grid is described, and a use case is presented.

Boundary voltage limits

Boundary voltage limits allow verifying voltage limit compliance within the elements represented by lumped models. The upper boundary voltage limits of producers, storages and Link-Grids are denoted as $\overline{BVL}_t^{\#-Pr}$, $\overline{BVL}_t^{\#-St}$, and $\overline{BVL}_t^{\#-\#}$, respectively, while the lower ones as $\underline{BVL}_t^{\#-Pr}$, $\underline{BVL}_t^{\#-St}$, and $\underline{BVL}_t^{\#-\#}$. The symbol # is replaced by the corresponding system levels, i.e. HV, MV, LV or CP.

The setup shown in Fig. 4.17 is analysed in detail to illustrate the variable BVLs of Link-Grids. Figure 4.17a shows the lumped model of an LV_Link-Grid, which is intended to be used for load flow analysis at the MV level. Its boundary node to the MV level is denoted as BLiN^{MV-LV} and is associated with the corresponding upper and lower boundary voltage limits, i.e. the $\overline{BVL}_t^{MV-LV}$ and $\underline{BVL}_t^{MV-LV}$. The lumped model represents the LV and CP_Link-Grids sketched in Fig. 4.17b. N residential CP_Link-Grids are connected equidistantly along the LV feeder with the length l through the corresponding boundary nodes, i.e. the BLiN^{LV-CP}. Due to the included consuming devices and PV systems, they absorb and inject power depending on daytime. The load profiles shown in Fig. 4.17c are used to represent the power contributions of each CP. ZIP models are used to consider the voltage dependency of consuming devices. The European grid codes set the corresponding limits, i.e. the $\overline{BVL}_t^{LV-CP}$ and $\underline{BVL}_t^{LV-CP}$, to $\pm 10\%$ around nominal voltage [25]. This specification makes the boundary voltage limits of CP_Link-Grids appear time-constant, Fig. 4.17d.

The voltage profiles of the cable feeder ($l = 1$ km, $N = 25$) at $t_1 = 12:10$ and $t_2 = 18:00$ are shown in Fig. 4.18 for MV-LV boundary voltages that provoke violations of the upper and lower LV-CP boundary voltage limits.

At 12:10, the $\overline{BVL}_{12:10}^{LV-CP}$ is violated by the backmost CP when the MV-LV boundary voltage exceeds 1.031 p.u., setting the $\overline{BVL}_{12:10}^{MV-LV}$ value accordingly, Fig. 4.18a and Eq. (4.21a). When reducing the MV-LV boundary voltage to 0.889 p.u., the $\underline{BVL}_{12:10}^{LV-CP}$ is violated by the foremost CP, Fig. 4.18b. In this case, the lower limit is less restrictive at the BLiN^{MV-LV} than at the BLiN^{LV-CP}, Eq. (4.21b).

$$\overline{BVL}_{12:10}^{MV-LV} = \overline{BVL}_{12:10}^{LV-CP} - \left|\Delta U_{12:10}^{upper}\right| = 1.031\,\text{p.u.} \qquad (4.21a)$$

$$\underline{BVL}_{12:10}^{MV-LV} = \underline{BVL}_{12:10}^{LV-CP} - \left|\Delta U_{12:10}^{lower}\right| = 0.889\,\text{p.u.} \qquad (4.21b)$$

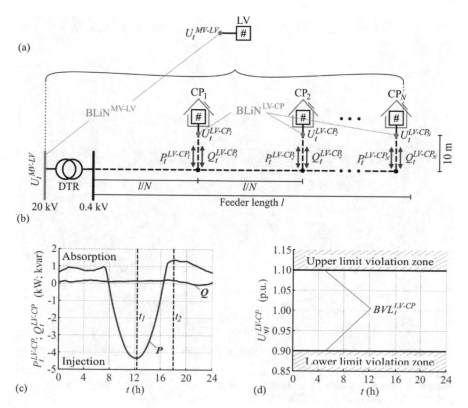

Fig. 4.17 Setup used to illustrate the occurrence of variable BVLs: **a** Lumped model of LV_Link-Grid; **b** LV_Link-Grid represented by the lumped model; **c** Power contributions of each CP; **d** Voltage limits at LV-CP boundary nodes

At 18:00, the voltages decrease along the DTR and LV feeder, setting the upper and lower MV-LV boundary voltage limits to 1.105 and 0.927 p.u., respectively, Fig. 4.18c–d and Eq. (4.22).

$$\overline{BVL}_{18:00}^{MV-LV} = \overline{BVL}_{18:00}^{LV-CP} + \left| \Delta U_{18:00}^{upper} \right| = 1.105 \, \text{p.u.} \tag{4.22a}$$

$$\underline{BVL}_{18:00}^{MV-LV} = \underline{BVL}_{18:00}^{LV-CP} + \left| \Delta U_{18:00}^{lower} \right| = 0.927 \, \text{p.u.} \tag{4.22b}$$

Calculating the load flows for the complete time horizon yields the variable upper and lower BVL_t^{MV-LV}-curves, Fig. 4.19.

The injection around noontime significantly decreases the upper BVL of the lumped LV_Link-Grid model and slightly decreases the lower one. Vice versa, in times of consumption, both BVL are increased, tightening the admissible voltage band in total. The extend of limit curve deformation—between the $BLiN^{MV-LV}$ and the $BLiN^{LV-CP}$—depends on the feeder parameters and the number of connected

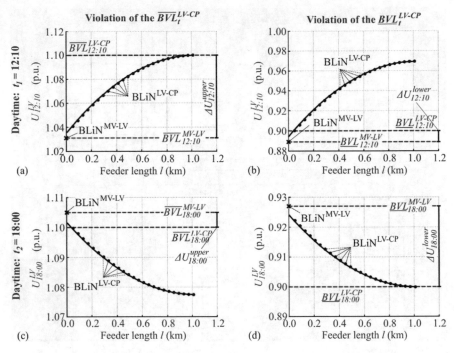

Fig. 4.18 Voltage profiles of the cable feeder ($l = 1$ km, $N = 25$) at t_1 and t_2 for MV-LV boundary voltages that provoke violations of the upper and lower LV-CP boundary voltage limits: **a** Upper limit at t_1; **b** Lower limit at t_1; **c** Upper limit at t_2; **d** Lower limit at t_2

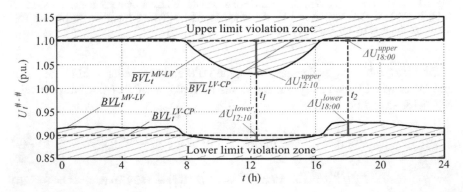

Fig. 4.19 Deformation of the boundary voltage limits by the internal voltage drops of the LV_Link-Grid (cable, $l = 1$ km, $N = 25$)

CPs. Figure 4.20 shows the upper and lower BVL_t^{MV-LV} for overhead line and cable conductors and for different feeder lengths and CP numbers.

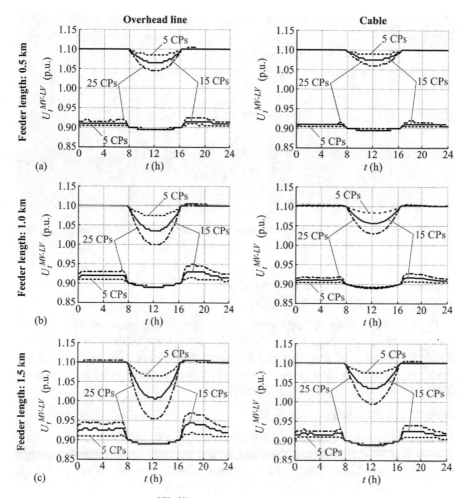

Fig. 4.20 Upper and lower BVL_t^{MV-LV} for 5, 15 and 25 connected CPs, for overhead line and cable conductors, and for different feeder lengths: **a** 0.5 km; **b** 1.0 km; **c** 1.5 km

The more CPs are connected, the tighter are the resulting BVL_t^{MV-LV}. The cable feeder is less restricted than the overhead line one, as it has lower series impedance. Long feeders are more restricted than short ones.

Components of the extended model

The *LINK*-based holistic architecture provides a systematic approach for power system modelling as it relies on the fractal feature of power systems (see Sect. 2.3). Figure 4.21 shows an overview of the extended monitoring model. The Link-Grid under study may apply to HV, MV, LV or CP_Link-Grids, or even to the conductors in consuming device level, Fig. 4.21a. Neighbouring Link-Grids are represented by the extended monitoring model, which complements the conventional one by two additional variables:

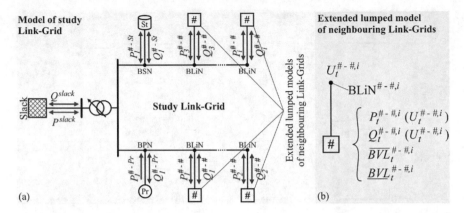

Fig. 4.21 Overview of the extended monitoring model: **a** Detailed model of the study Link-Grid; **b** Lumped model of neighbouring Link-Grids

- upper boundary voltage limit ($\overline{BVL}$); and.
- lower boundary voltage limits ($\underline{BVL}$).

Figure 4.21b shows the extended lumped model supplemented with the new changing boundary voltage limits. The neighbouring Link-Grids are represented by the "$\boxed{\#}$" symbol to distinguish between the conventional and the extended lumped model. Producers and Storages are connected through the Boundary Producer Nodes (BPN) and the Boundary Storage Nodes (BSN).

Detailed model of the study Link-Grid

The detailed Link-Grid model includes all existent lines, transformers, reactive power devices, and HVDC systems interconnected and parametrised to represent the grid part under study. The model is complemented with voltage limits for each node, representing the electrical equipment's insulation-related limits.

Extended lumped models of the connected elements

The lumped model behind each *PQ* node-element is complemented with the boundary voltage limits so that it includes the following information for each regarded instant of time:

- The **power contributions** as functions of the boundary voltage, i.e. $P_t(U_t)$ and $Q_t(U_t)$;
- And the **boundary voltage limits**, i.e. upper and lower BVL_t.

The **lumped producer models** reflect the behaviour of electricity production facilities. They inject active power and inject or absorb reactive power. Their boundary voltage limits, $BVL_t^{\#-Pr}$, are usually time-invariant and documented in the datasheet provided by the manufacturer.

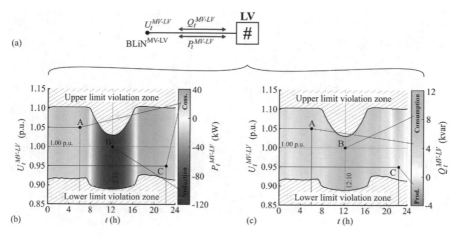

Fig. 4.22 Exemplary lumped LV_Link-Grid model: **a** Symbolic representation; **b** Active power contribution and BVL_t^{MV-LV}; **c** Reactive power contribution and BVL_t^{MV-LV}

The **lumped storage models** reflect the behaviour of storage facilities. They inject or absorb active and reactive power. Their boundary voltage limits, $BVL_t^{\#-St}$, are usually time-invariant and documented in the datasheet provided by the manufacturer.

The **lumped Link-Grid models** reflect the aggregate behaviour of neighbour Link-Grids and all thereto connected elements, i.e. producers, storages and further Link-Grids. They inject or absorb active and reactive power. Their boundary voltage limits, $BVL_t^{\#-\#}$, usually vary over time.

Figure 4.22 exemplifies the lumped Link-Grid model using the LV_Link-Grid described in Fig. 4.17 (cable, $l = 1$ km, $N = 25$). The power contributions of the lumped model are defined for each instant of time and boundary voltage value. Each point (t, U) within the violation zones provokes voltage limit violations within the Link-Grid represented by the lumped model.

The active power behaviour is presented in the relevant zone between the upper and lower BVL_t^{MV-LV} using different colour shades from red for 120 kW injection to light violet for 40 kW absorption, Fig. 4.22b. The direction of the active power flow changes twice a day. From 00:00 to 8:00, the LV_Link-Grid draws active power from the MV level. At 6:00, for a boundary voltage of 1.05 p.u., it consumes 21.48 kW, case A. With the onset of a sunny day, PV systems begin to produce electricity. When the total production exceeds the total consumption, i.e. from 08:00 to 16:15, the active power changes the direction and flows from the LV to the MV level. It rises steadily to reach the maximum value at 12:10. For a voltage of 1.00 p.u. at the BLiN^{MV-LV}, the LV_Link-Grid injects 101.93 kW into the MV level, case B. With the offset of a sunny day, the electricity production of PV systems reduces to zero. When the produced electricity falls below the consumed one, i.e. from 16:15 to 24:00, the active power changes the direction and flows again from the MV to the LV level. At 22:10, for a boundary voltage of 0.95 p.u., the lumped model consumes 21.94 kW, case C. Figure 4.22c shows the reactive power behaviour in different colour

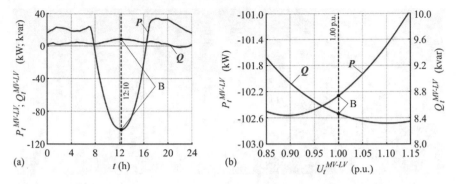

Fig. 4.23 Power contributions of the exemplary lumped Link-Grid model: **a** As functions of time for the boundary voltage of 1.00 p.u.; **b** As functions of boundary voltage at 12:10

shades from light blue for 4 kvar injection to cyan for 12 kvar consumption. Also, the reactive power changes the direction of flow twice a day. It flows from the MV into the LV level from 0:00 to 20:45, reaching values of 4.51 and 8.46 kvar in cases A and B, respectively. Later on, due to the capacitive nature of many modern consuming devices (mostly LED light bulbs), the reactive power changes the flow direction: It flows from the LV into the MV level until 23:30. Here, the boundary voltage has a noticeable impact on reactive power production. The lower the boundary voltage, the less reactive power is produced, amounting to 0.97 kvar in case C.

Figure 4.23a, b show the power contributions for one specific boundary voltage and instant of time, respectively, cut out from the lumped model shown in Fig. 4.22. Case B is marked in both figures.

The time dependency of active and reactive power contributions appears similar to the one of the CPs shown in Fig. 4.17c, as only grid losses are added, Fig. 4.23a. The power contributions at 12:10 only slightly depend on the boundary voltage, Fig. 4.23b. Both the active power injection and the reactive power absorption decrease with increasing voltage.

Calculation procedure of the model components

The aggregate behaviour of any Link-Grid can be calculated if its structure and the behaviour of the connected elements are known. Figure 4.24 illustrates the aggregation of the Link-Grid. The Boundary Link Node (BLiN) at which the Link-Grid is aggregated is designated as the "BLiN of aggregation".

The Link-Grid to be aggregated connects producers, storages, and neighbouring Link-Grids, Fig. 4.24a. As described in Sect. 4.6.1.3, their lumped models include the power contributions as functions of boundary voltage and time and the corresponding boundary voltage limits. Figure 4.24b shows the aggregated Link-Grid, which is calculated by using the procedures described below.

1   Connect the slack node-element to the BLiN of aggregation.
2   Define the slack voltage range of interest, e.g. from 0.9 p.u. to 1.1 p.u., and the corresponding resolution, e.g. 0.01 p.u. steps.

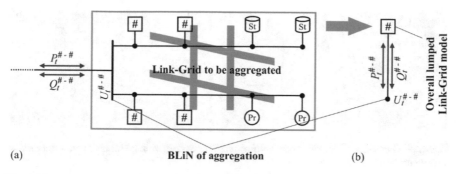

Fig. 4.24 Aggregation of a Link-Grid: **a** Link-Grid to be aggregated; **b** Overall lumped Link-Grid model

3 Select one instant of time specified by the lumped models of the connected elements.

4 Repeat load flow simulations of the selected instant of time for all defined slack voltages, and record the P- and Q-values provided by the slack node-element. Furthermore, document all slack voltage values that provoke violations of:

- The voltage limits of any node of the detailed Link-Grid model, i.e. the Link-Grid to be aggregated.
- And the boundary voltage limits of any connected element.

5 Repeat Steps 1–4 for All Other Instants of Time Specified by the Lumped Models of the Connected Elements.

This procedure yields the following results for each instant of time:

- The power contributions of the Link-Grid to be aggregated as functions of its boundary voltage;
- A set of slack voltage values that provoke violations of upper voltage limits either within the Link-Grid to be aggregated; or at the boundary node of any connected lumped model. This set of values is denoted as $U_{k,t}^{upper-viol.}$, where k indexes the different values within this set;
- And a set of slack voltage values that provoke violations of lower voltage limits either within the Link-Grid to be aggregated; Or at the boundary node of any connected lumped model. This set of values is denoted as $U_{m,t}^{lower-viol.}$, where m indexes the different values within this set.

Subsequently, the upper and lower boundary voltage limits of the lumped Link-Grid model are calculated using Eq. (4.23).

$$\overline{BVL}_t^{\#-\#} = \min_k \left(U_{k,t}^{upper-viol.} \right) \tag{4.23a}$$

$$\underline{BVL}_t^{\#-\#} = \max_m \left(U_{m,t}^{lower-viol.} \right) \tag{4.23b}$$

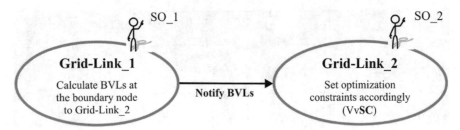

Fig. 4.25 Overview of the use case: Notification of boundary voltage limits when different SOs operate the Grid-Links

The calculated power contributions and boundary voltage limits build the overall lumped Link-Grid model, which can be used to study the behaviour of the Link-Grid connected at the other side of the BLiN of aggregation.

Use case: Notification of boundary voltage limits
The extended monitoring model allows formulating use cases for the *LINK*-based monitoring chain process (see Sect. 2.6.1) that apply in different timeframes, e.g. in day-ahead and real-time [59].

Generalised use case
Due to the standardised structure of the *LINK*-based holistic architecture, the consideration of two Grid-Links is sufficient to formulate the generalised use case. Two cases are distinguished: Different System Operators (SO) operate the Grid-Links, and one SO operates the Grid-Links (Fig. 4.25).

- Different operators operate the Grid-Links:
  This case is shown in Fig. 4.25, where SO_1 and SO_2, which may be different companies, operate Grid-Link_1 and Grid-Link_2, respectively. SO_1 calculates the BVLs at the boundary node to Grid-Link_2 and notifies them to SO_2. SO_2 sets the optimisation constraints[3] of its Volt/var Secondary Control (*VvSC*) accordingly.

- The same operator operates the Grid-Links:
  Figure 4.26 shows the case where one SO, e.g. the DSO, operates both Grid-Links. Although the data of both Grid-Links are normally available in the same database, the state estimation may be performed by different applications and in different time intervals. The application estimating the state of Grid-Link_1 calculates the BVLs at the boundary node to Grid-Link_2 and notifies them to the application observing Grid-Link_2. The *VvSC* in Grid-Link_2 adapts its optimisation constraints accordingly.

---

[3] These constraints apply exclusively to the boundary node between Grid-Link_1 and Grid-Link_2.

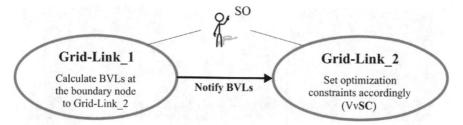

Fig. 4.26 Overview of the use case: Notification of boundary voltage limits when one SO operates the Grid-Links

Day-ahead scheduling

SO_1 calculates the day-ahead schedule of the BVLs at the boundary node of Grid-Link_2 with sufficient resolution and notifies it to SO_2. SO_2 sets the (time-variant) optimisation constraints of its VvSC for the next day accordingly.

Short-term adaptation

SO_1 recognises a deviation from its actual day-ahead schedule of the BVLs at the boundary node to Grid-Link_2, recalculates the corresponding BVLs and notifies them to SO_2. SO_2 updates the optimisation constraints of its VvSC accordingly.

4.6.1.4 Conventional Versus Extended Model

The use of time constant voltage limits to check the power flow results is not accurate. Voltage limits in radial structures (e.g. MV) are object to deformation due to the voltage drops in the subordinate grids (e.g. LV). The extent of this deformation depends on the properties of the LV feeders: The greater the feeder impedance and the number of connected customer plants, the more intensive is the deformation. In order to take the latter into account, the conventional lumped models of grid parts are extended with new parameters, i.e. the upper and lower boundary voltage limits. BVLs allow verifying voltage limit compliance within the study/operation Link-Grid and within the neighbouring Link-Grids, represented by lumped models in power flow analysis.

The presented use cases allow increasing the operational efficiency of Smart Grids by day-ahead scheduling and short-term (online) adaption of the boundary voltage limits.

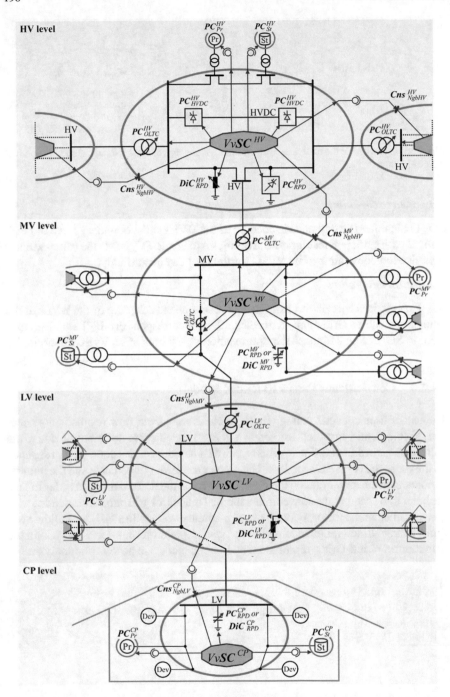

Fig. 4.27 Overview of the generalised Volt/var chain control from the holistic view of the power system

4.6.2 General Description of the Volt/var Chain

The Volt/var chain control coordinates the available control variables using the secondary control (see Sect. 2.5.6.2) as the base instrument. Figure 4.27 shows the generalised form of the Volt/var chain control wherein the Grid-Links are set upon four classical levels: HV, MV, LV and CP level. But, by definition, the Grid-Link size is variable (see Sect. 2.5.2.3). It may apply to the classical system levels and to a part of the grid that includes more than one system level, e.g. MV and LV.

The LV and CP levels are distinguished to enable the consideration of different ownerships and stakeholder interests. While the automation and communication path is drawn in blue, the power flow path is black coloured.

In its generalised form, the Volt/var chain control utilises all reactive power resources, including storages with reactive power capabilities across all system levels. The neighbour Grid-Links may act as additional control variables by accepting reactive power set-points. The $VvSCs$ of the generalised Volt/var chain control are described in the following sections separately for the horizontal and vertical power system axis.

4.6.2.1 Horizontal

The transmission grids, i.e. the HV grids, lie within the horizontal axis (X-axis) of the power system. HV_Grid-Links are typically set upon the control areas of the TSOs. Equation (4.24) compactly represents the control variables and dynamic constraints of the horizontal Volt/var chain control.

$$VvSC_{Chain}^{X-axis} = \bigcup_{i=1}^{M} \left\{ VvSC^{HV_i}(PC_{OLTC}^{HV_i}, \ PC_{Pr}^{HV_i}, \ PC_{St}^{HV_i}, \ PC_{RPD}^{HV_i}, \ DiC_{RPD}^{HV_i}, \right.$$

$$\left. PC_{HVDC}^{HV_i}, \ SC_{NgbHV}^{HV_i}, \ SC_{NgbMV}^{HV_i}; \ Cns_{NgbHV}^{HV_i}) \right\} \tag{4.24}$$

where the $VvSC^{HV_i}$ calculates in real-time:

- The voltage set-points for the primary controls *PC OLTC HV_i* of the power transformers included in the HV_Grid-Link i that have OLTC;
- The voltage and reactive power set-points for the primary controls *PC PrHV_i* of the Producer-Links connected to the HV_Grid-Link i;
- The voltage and reactive power set-points for the primary controls *PC StHV_i* of the Storage-Links connected to the HV_Grid-Link i;
- The voltage, reactive power and switch position set-points for the primary *PC RPDHV$_i$* and direct controls *DiC RPDHV$_i$* of the RPDs included in the HV_Grid-Link i;
- The voltage and reactive power set-points for the primary controls *PC HVDC HV_i* of the HVDC converters included in the HV_Grid-Link i; and.
- The reactive power set-points for the secondary controls $SC_{NgbHV}^{HV_i}$ of the neighbouring HV_Grid-Links that act as var control variables.

- And the reactive power set-points for the secondary controls $SC_{NgbMV}^{HV_i}$ of the neighbouring MV_Grid-Links;

While respecting:

- The reactive power constraints $Cns_{NgbHV}^{HV_i}$ at the boundary nodes to the neighbouring HV_Grid-Links that act as var constraints.

4.6.2.2   Vertical

The vertical power system axis (Y-axis) comprises all system levels, i.e. HV, MV, LV and CP level. Equation (4.25) compactly represents the control variables and dynamic constraints of the vertical Volt/var chain control.

$$
\begin{aligned}
VvSC_{Chain}^{Y-axis} = \big\{ & VvSC^{HV}(PC_{OLTC}^{HV}, PC_{Pr}^{HV}, PC_{St}^{HV}, PC_{RPD}^{HV}, \\
& DiC_{RPD}^{HV}, PC_{HVDC}^{HV}, SC_{NgbHV}^{HV}, SC_{NgbMV}^{HV}; Cns_{NgbHV}^{HV}), \\
& VvSC^{MV}\Big(PC_{OLTC}^{MV}, PC_{Pr}^{MV}, PC_{St}^{MV}, PC_{RPD}^{MV}, DiC_{RPD}^{MV}, SC_{NgbLV}^{MV}; Cns_{NgbHV}^{MV}\Big), \\
& VvSC^{LV}\Big(PC_{OLTC}^{LV}, PC_{Pr}^{LV}, PC_{St}^{LV}, PC_{RPD}^{LV}, DiC_{RPD}^{LV}, SC_{NgbCP}^{LV}; Cns_{NgbMV}^{LV}\Big), \\
& VvSC^{CP}\Big(PC_{Pr}^{CP}, PC_{St}^{CP}, PC_{RPD}^{CP}, DiC_{RPD}^{CP}; Cns_{NgbLV}^{CP}\Big)\big\}
\end{aligned}
\tag{4.25}
$$

HV level
The $VvSC^{HV}$ calculates in real-time:

- The voltage set-points for the primary controls PC_{OLTC}^{HV} of the power transformers included in the HV_Grid-Link that have OLTC;
- The voltage and reactive power set-points for the primary controls PC_{Pr}^{HV} of the Producer-Links connected to the HV_Grid-Link;
- The voltage and reactive power set-points for the primary controls PC_{St}^{HV} of the Storage-Links connected to the HV_Grid-Link;
- The voltage, reactive power and switch position set-points for the primary PC_{RPD}^{HV} and direct controls DiC_{RPD}^{HV} of the RPDs included in the HV_Grid-Link;
- The voltage and reactive power set-points for the primary controls PC_{HVDC}^{HV} of the HVDC converters included in the HV_Grid-Link;
- The reactive power set-points for the secondary controls SC_{NgbHV}^{HV} of the neighbouring HV_Grid-Links that act as var control variables;
- And the reactive power set-points for the secondary controls SC_{NgbMV}^{HV} of the neighbouring MV_Grid-Links;

While respecting:

- The reactive power constraints Cns_{NgbHV}^{HV} at the boundary nodes to the neighbouring HV_Grid-Links that act as var constraints.

MV level

The $VvSC^{MV}$ calculates in real-time:

- The voltage set-points for the primary controls PC_{OLTC}^{MV} of the supplying transformers and other transformers included in the MV_Grid-Link that have OLTC;
- The voltage and reactive power set-points for the primary controls PC_{Pr}^{MV} of the Producer-Links connected to the MV_Grid-Link;
- The voltage and reactive power set-points for the primary controls PC_{St}^{MV} of the Storage-Links connected to the MV_Grid-Link;
- The voltage, reactive power and switch position set-points for the primary PC_{RPD}^{MV} and direct controls DiC_{RPD}^{MV} of the RPDs included in the MV_Grid-Link;
- And the reactive power set-points for the secondary controls SC_{NgbLV}^{MV} of the neighbouring LV_Grid-Links;

  While respecting:

- The reactive power constraints Cns_{NgbHV}^{MV} at the boundary node to the neighbouring HV_Grid-Link.

LV level

The $VvSC^{LV}$ calculates in real-time:

- The voltage set-points for the primary control PC_{OLTC}^{LV} of the distribution transformer included in the LV_Grid-Link (when it possesses an OLTC);
- The voltage and reactive power set-points for the primary controls PC_{Pr}^{LV} of the Producer-Links connected to the LV_Grid-Link;
- The voltage and reactive power set-points for the primary controls PC_{St}^{LV} of the Storage-Links connected to the LV_Grid-Link;
- The voltage, reactive power and switch position set-points for the primary PC_{RPD}^{LV} and direct controls DiC_{RPD}^{LV} of the RPDs included in the LV_Grid-Link;
- And the reactive power set-points for the secondary controls SC_{NgbCP}^{LV} of the neighbouring CP_Grid-Links;

  While respecting:

- The reactive power constraints Cns_{NgbMV}^{LV} at the boundary node to the neighbouring MV_Grid-Link.

CP level

The $VvSC^{CP}$ calculates in real-time:

- The voltage and reactive power set-points for the primary controls PC_{Pr}^{CP} of the Producer-Links connected to the CP_Grid-Link;
- The voltage and reactive power set-points for the primary controls PC_{St}^{CP} of the Storage-Links connected to the CP_Grid-Link;
- And the switch position set-points for the primary PC_{RPD}^{CP} and direct controls DiC_{RPD}^{CP} of the RPDs[4] included in the CP_Grid-Link;

---

[4] RDPs are commonly used in industrial CPs.

While respecting:

- The reactive power constraint $\mathbf{Cns}^{CP}_{NgbLV}$ at the boundary node to the neighbouring LV_Grid-Link.

4.6.3   New Volt/var Control Strategy in LV and CP Level

The *LINK*-Architecture stipulates that each Grid-Link operator should mainly use its own control devices to maintain acceptable voltages and power factor. Reactive power support from neighbour elements, such as Grid-, Producer- and Storage-Links, may be incorporated into the Volt/var control process when the internal resources are insufficient and when their installation is uneconomical. This idea gave rise to a new Volt/var control strategy the $X(U)$-control.

4.6.3.1   Distributed and Concentrated Var Contributions

The categorisation of the shunt var contribution into the distributed and concentrated type is introduced in [35]. Here, the associated effectiveness is analysed in two steps: Firstly, an unloaded single-phase feeder is used to study the fundamental impact of both var contribution types on the system behaviour,Several simplifications are made to enable the analytical investigation in closed-form. Secondly, simulations are conducted on the feeder in loaded conditions to validate the findings.

Definitions
The distributed and concentrated var contributions provoke distinct reactive power flows that differently affect the grid voltages.

Distributed var contributions

> Per definition, the distributed var contributions are reactive power contributions at various nodes distributed throughout the length of the feeder, Fig. 4.28.

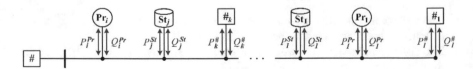

Fig. 4.28   Overview of the distributed var contributions

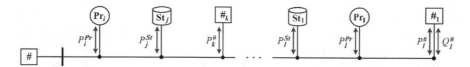

Fig. 4.29 Overview of the concentrated var contribution

Concentrated var control.

> Per definition, the concentrated var contribution is one reactive power contribution close to the end of the feeder, Fig. 4.29.

Behaviour of the unloaded feeder

The fundamental impact of both var contribution types on the voltage is analysed using an unloaded feeder, i.e. a feeder through which no active power flows. The feeder consists of N line segments with the reactance X, $(N + 1)$ nodes with the voltage U_i, and N connected CPs, Figs. 4.30 and 4.31. The active and reactive power losses of the line segments are neglected, and the node voltages are calculated with Eq. (4.12) to enable the analysis of the feeder behaviour in closed-form.

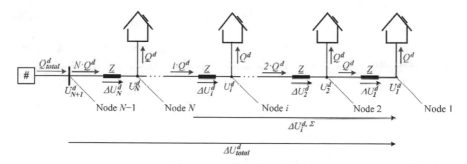

Fig. 4.30 Setup used to analyse the unloaded feeder's behaviour in the presence of distributed var contributions

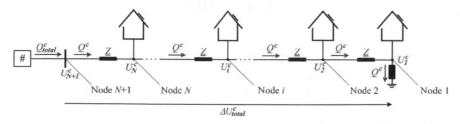

Fig. 4.31 Setup used to analyse the unloaded feeder's behaviour in the presence of one concentrated var contribution

Distributed var contributions

The unloaded feeder's behaviour in the presence of distributed var contributions is analysed using the setup shown in Fig. 4.30. The CPs connected at nodes $i \in [1; N]$ provoke distributed var contributions[5] Q^d. The Link-Grid connected at node $N + 1$ contributes the reactive power Q^d_{total} required to satisfy the overall reactive power balance.

The voltage drop ΔU^d_k over the line segment k depends on its reactance, the reactive power flow $(k \cdot Q^d)$ through it, and the voltage of the corresponding node U^d_k, Eq. (4.26).

$$\Delta U^d_k = \frac{(k \cdot Q^d) \cdot X}{U^d_k} \tag{4.26}$$

The voltage drop $\Delta U^{d,\Sigma}_k$ between node k and feeder end results from adding up the voltage drops over the intermediate line segments, Eq. (4.27).

$$\Delta U^{d,\Sigma}_k = \sum_{i=1}^{k} \Delta U^d_i \tag{4.27}$$

For $k \in [2; N + 1]$, the voltage U^d_k at node k is expressed by the voltage U^d_1 at feeder end and the voltage drop in between, Eq. (4.28).

$$U_k = U^d_1 + \Delta U^{d,\Sigma}_{k-1} \tag{4.28}$$

Combining Eqs. (4.26)–(4.28) yields the following recursive formula for the total voltage drop ΔU^d_{total} over the feeder with N line segments and distributed var contributions.

$$\Delta U^d_{total} = \Delta U^{d,\Sigma}_N = \frac{Q^d \cdot X}{U^d_1} + \sum_{i=2}^{N} \frac{i \cdot Q^d \cdot X}{U^d_1 + \Delta U^{d,\Sigma}_{i-1}} \tag{4.29}$$

As the losses are neglected, the total reactive power that flows into the feeder is given by Eq. (4.30).

$$Q^d_{total} = N \cdot Q^d \tag{4.30}$$

Concentrated var contribution

The behaviour of the unloaded feeder is also analysed for the concentrated var contribution, Fig. 4.31. While the CPs connected at nodes $i \in [1; N]$ do not contribute any reactive power, the RPD connected at node $i = 1$ contributes[6] the reactive power

---

[5] Positive algebraic sign for reactive power absorptions.

Q^c. As in the case with distributed var contributions, the Link-Grid connected at node $N + 1$ satisfies the overall reactive power balance.

The total voltage drop ΔU_{total}^c over the feeder depends on the concentrated reactive power contribution Q^c, the feeder reactance $(N \cdot X)$ and the voltage at feeder end, Eq. (4.31).

$$\Delta U_{total}^c = \frac{Q^c \cdot (N \cdot X)}{U_1^c} \tag{4.31}$$

The voltage of each node k is given by Eq. (4.32).

$$U_k^c = U_1^c + \frac{Q^c \cdot (k - 1) \cdot X}{U_1^c} \tag{4.32}$$

As losses are neglected, the total reactive power that flows into the feeder equals the concentrated reactive power contribution, Eq. (4.33).

$$Q_{total}^c = Q^c \tag{4.33}$$

Distributed versus concentrated var contributions

The var contribution types are compared by setting the same voltages at feeder end and feeder beginning as in Eq. (4.34).

$$U_1^d = U_1^c = U_1 \tag{4.34a}$$

$$\Delta U_{total}^d - \Delta U_{total}^c = \Delta U_{total}^{d/c} \tag{4.34b}$$

Combining Eqs. (4.29), (4.31) and (4.34) yields the concentrated reactive power contribution that is required to achieve the same total voltage drop as the distributed ones as a function of the line segment number, Eq. (4.35).

$$Q^c = \frac{Q^d}{N} \cdot \left(1 + \sum_{i=2}^{N} i \cdot \frac{U_1}{U_1 + \Delta U_{i-1}^{d, \Sigma}}\right) \tag{4.35}$$

Equations (4.30)–(4.35) allow calculating the total reactive power flowing into the feeder for the distributed and concentrated var contributions and the total voltage drop over the feeder, depending on the line segment number. The values used for the voltage at the feeder end and for the distributed var contributions are given in Table 4.3.

The resulting curves are shown in Fig. 4.32a, b for exemplary LV feeder parameters[6] and for var absorptions and injections, respectively.

---

[6] The following parameters that correspond to a LV overhead line with a length of 100 m are used: $U_{nom} = 230$ V, $R = 0.03264$ Ω, $X = 0.03557$ Ω.

Table 4.3 Parameters used to analyse the behaviour of the unloaded feeder

Var contribution	U_1 (p.u.)	Q^d (kvar)
Absorption	0.9	0.5
Injection	1.1	− 0.5

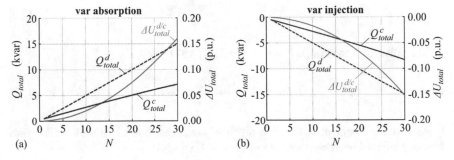

Fig. 4.32 Total voltage drop and reactive power requirement of the unloaded feeder for distributed and concentrated var contributions as functions of the line segment number: **a** Var absorption case; **b** Var injection case

The distributed var contributions require more reactive power in total than the concentrated one to achieve the same total voltage drop over the feeder. For example, regarding a feeder with 20 line segments and the var absorption case, the distributed var contributions provoke a total voltage drop and reactive power requirement of 0.076 p.u. and 10 kvar, respectively. Meanwhile, the concentrated var contribution requires only 5.05 kvar to achieve the same total voltage drop.

The unloaded feeder's voltage profiles resulting from the distributed and concentrated var contributions are calculated according to Eqs. (4.28) and (4.32) for the absorption and injection cases. They are shown in Fig. 4.33 for the feeder with 20 line segments. The voltage profile has a curved shape when the var contributions are distributed, and a straight one when one concentrated var contribution is present. In the var absorption case, the total voltage drop of 0.076 p.u. appears as discussed

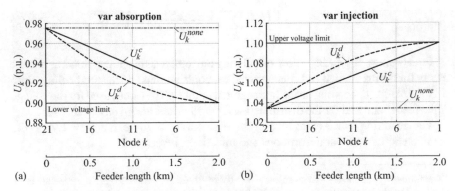

Fig. 4.33 The unloaded feeder's voltage profiles for no, distributed and concentrated var contributions: **a** Var absorption case; **b** Var injection case

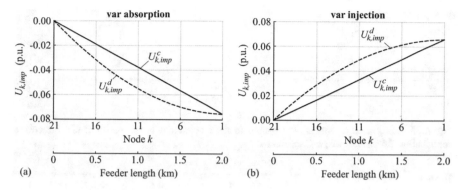

Fig. 4.34 Impact of the distributed and concentrated var contributions on the unloaded feeder's voltage profile:**a** Var absorption case;**b** Var injection case

above. The node voltages U_k^{none} without var contributions are constant along the feeder, as no active and reactive power flows through the line segments.

The impact of both var contribution types is calculated by Eq. (4.36).

$$U_{k,imp}^d = U_k^d - U_k^{none} \tag{4.36a}$$

$$U_{k,imp}^c = U_k^c - U_k^{none} \tag{4.36b}$$

The curves describing the var contributions' impact have the same shape as the corresponding voltage profiles due to the constant voltages along the feeder without var contributions, Fig. 4.34. The distributed and concentrated var absorptions decrease the node voltages along the feeder, while the var injections increase them. The largest impact appears at the feeder end in all cases.

The closed-form analysis of the unloaded LV feeder revealed the clear superiority of the concentrated var contribution as follows:

> The concentrated var contribution provokes lower reactive power flows at the LV feeder beginning than the distributed ones to achieve the same total voltage drop over the feeder.

Behaviour of the loaded feeder

To verify the previous section's findings under more realistic conditions, the feeder behaviour is also analysed for loaded conditions. Therefore, load flow simulations are conducted for the var absorption and injection cases using the setups[7] shown in Fig. 4.35. Distributed and concentrated var contributions, as well as the case without any var contributions, are simulated.

---

[7] Positive algebraic sign for active power injection and reactive power absorption.

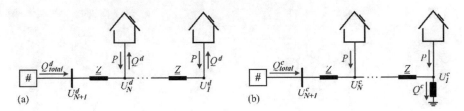

Fig. 4.35 Setup used to analyse the loaded feeder's behaviour in the presence of different var contributions: **a** Distributed; **b** Concentrated

Table 4.4 Parameters used to analyse the behaviour of the loaded feeder	Var contribution	U_{N+1} (p.u.)	Q^d (kvar)	P (kW)
	Absorption	1.0	0.5	0.7
	Injection	1.0	-0.5	-0.7

The values used for the voltage at feeder beginning and for the distributed active and reactive power contributions are given in Table 4.4. In analogy with the analysis of the unloaded feeder, the concentrated reactive power contribution is always set to achieve the same total voltage drop as the distributed one, Eq. (4.34b).

Both the active and reactive power flows determine the behaviour of the loaded feeder. The impacts of both var contribution types are isolated according to Eqs. (4.37) and (4.38) to study their effects on the feeder behaviour.

$$\Delta U_{imp}^{d/c} = \Delta U_{total}^{d/c} - \Delta U_{total}^{none} \tag{4.37}$$

$$Q_{imp}^{d} = Q_{total}^{d} - Q_{total}^{none} \tag{4.38a}$$

$$Q_{imp}^{c} = Q_{total}^{c} - Q_{total}^{none} \tag{4.38b}$$

where ΔU_{total}^{none} and Q_{total}^{none} are the total voltage drop and reactive power requirement at feeder beginning, respectively, when no reactive power is contributed by any element connected along the feeder. The corresponding curves are shown in Fig. 4.36a, b for the var absorption and injection cases, respectively.

Fundamentally, the same behaviour is observed as for the unloaded feeder (see Fig. 4.32). The distributed var contributions always require more reactive power than the concentrated one, especially for high line segment numbers.

The loaded feeder's voltage profiles for no, distributed and concentrated var contributions are shown in Fig. 4.37a, b for the var absorption and injection case, respectively. The cases without any var contributions along the feeder clearly show the impact of the active power flows. The injections and absorptions of active power increase and decrease the node voltages, respectively. var contributions allow mitigating the voltage change caused by the active power flows. In comparison, the distributed var contributions cause relatively flat voltage profiles so that the maximum and minimum voltages

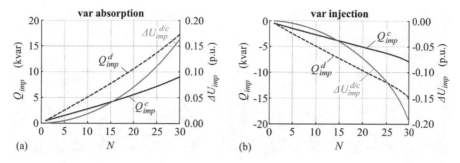

Fig. 4.36 Impact of the distributed and concentrated var contributions on the total voltage drop and reactive power requirement of the loaded feeder as functions of the line segment number: **a** Var absorption case; **b** Var injection case

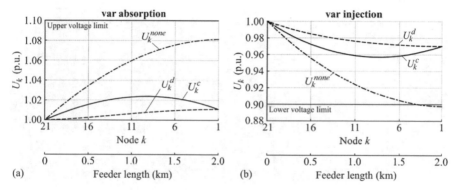

Fig. 4.37 The loaded feeder's voltage profile for no, distributed and concentrated var contributions: **a** Var absorption case; **b** Var injection case

appear at the feeder end for the var absorption and injection cases, respectively. Meanwhile, the concentrated var contribution provokes more curved profiles, wherein the maximum and minimum voltages appear somewhere along the feeder.

The curves describing the impact of the distributed and concentrated var contributions on the loaded feeder's voltage profile are quite similar to those of the unloaded feeder, Fig. 4.38.

> The simulations of the loaded feeder have verified the conclusions drawn from the closed-form analysis of the unloaded feeder.

4.6.3.2 $X(U)$-Control

The term '$X(U)$' refers to a voltage-dependent reactance that adjusts itself to inject or absorb the reactive power amount required to maintain acceptable voltages in LV

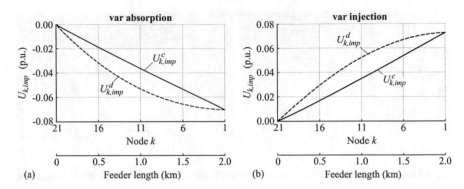

Fig. 4.38 Impact of the distributed and concentrated var contributions on the loaded feeder's voltage profile: **a** Var absorption case; **b** Var injection case

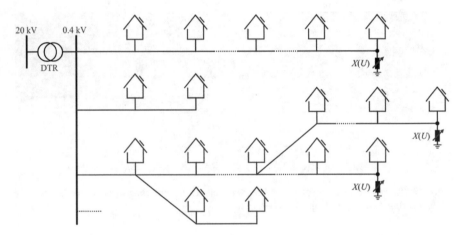

Fig. 4.39 Illustration of the $X(U)$-control strategy

level. In contrast to the $L(U)$-control [33], which can only absorb reactive power, the $X(U)$-control is capable of absorbing and injecting reactive power to mitigate violations of the upper and lower voltage limits. Figure 4.39 illustrates the concept of the $X(U)$-control: one controllable shunt-connected RPD is connected close to the end of each violating feeder, i.e. each feeder that may violate the BVLs of thereto connected elements (mainly customer plants). More than one RPD may be necessary when the feeder is branched.

Mechanically switched capacitors and reactors, as well as static var compensators may be used for this shunt reactance insertion (see Table 4.2). As an alternative, the $X(U)$-control may also be realised by shunt current injections, e.g. by using STATCOMs.

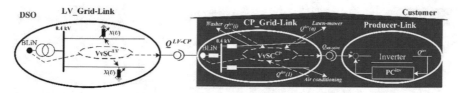

Fig. 4.40 Overview of the Volt/var chain control in LV and CP level when $X(U)$ primary control is used

Two types of $X(U)$-control are generally distinguished: Primary and local (see Sect. 4.2.5). In any case, its use resolves the intertwining of the LV grid and CP operation provoked by the $Q(U)$-control (see Sect. 2.10.2 and Sect. 4.5).

$X(U)$ primary control

Figure 4.40 shows an overview of the Volt/var chain control in LV and CP level when the $X(U)$ primary control is used. The $VvSC^{LV}$ adapts the voltage set-points of the $X(U)$-control devices' primary controls and the reactive power set-points to be respected by the $VvSC^{CP}$ of the connected CPs. The $VvSC^{CP}$ adapts the primary control settings of the PV inverter to meet the constraint at the $BLiN^{LV-CP}$.

$X(U)$ local control

The $X(U)$ local control with fixed control parameters may be used as an alternative to the combination of primary and secondary controls in LV level. In this case, the RPDs inject or absorb reactive power to maintain their terminal voltages within a predefined band, e.g. between 0.91 and 1.09 p.u.

4.6.3.3   CP_Q-Autarky

CP_Q-Autarky is a special case of the $VvSC$ realised at the CP level where the reactive power constraint between LV and CP level is set to zero. This means the concept of load compensation, which is traditionally used at the device level or in industrial CPs (see Sect. 4.2.3.1), is applied to all types of CPs, including residential ones. The reactive power needed for the functioning of rotating devices such as washing machines, lawnmowers, and air conditioning units is compensated directly at the CP level.

> Q-**Autarkic** or Q-**self-sufficient customer plants** produce the required reactive power within their own premises. Per definition, they do not exchange any reactive power with the LV_Grid-Link. In this case, the LV_Grid-Link serves prosumers and consumers by a factor of unity.

To realise CP_Q-Autarky, the $VvSC^{CP}$ adapts the control settings of the RPDs, Producer- and Storage-Links connected at the CP level to eliminate the reactive power flow through the LV-CP boundary node at all times, Fig. 4.41.

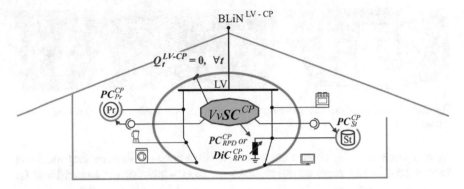

Fig. 4.41 Illustration of the CP_Q-Autarky

The compact representation of the control variables and dynamic constraints of Q-Autarkic CPs is given in Eq. (4.39).

$$VvSC^{CP}\left(PC_{Pr}^{CP}, PC_{St}^{CP}, PC_{RPD}^{CP}, DiC_{RPD}^{CP}; Cns_{NgbLV}^{CP} = 0 \text{ kvar}\right) \tag{4.39}$$

4.6.3.4 $X(U)$ and CP_Q-Autarky Control Ensemble

The presented Volt/var control strategies may be combined. Figure 4.42 shows the reactive power flows through an LV feeder when no control is used and when $X(U)$-control and its combination with CP_Q-Autarky are applied. The cyan coloured arrows represent the reactive power contributions of the consuming devices and the wires at the CP level. While the consuming devices contribute reactive power when used, the wires produce and consume reactive power depending on the voltage and the power flowing through them. The reactive power contributions of the control devices are shown in blue colour.

When no control is applied, the CPs exchange reactive power with the LV_Link-Grid, provoking distributed reactive power contributions along the feeder, Fig. 4.42a. The use of $X(U)$-control provokes—in addition to the distributed CP contributions – one concentrated reactive power contribution close to the end of the LV feeder, Fig. 4.42b. The combination of $X(U)$-control with CP_Q-Autarky eliminates the distributed reactive power contributions, Fig. 4.42c.

4.7 Link-Grids' Volt/var Behaviour Using Different Control Strategies

This section systematically analyses the MV, LV and CP_Link-Grids' behaviour using different Volt/var control strategies and for the case without any Volt/var control [61]. Different real Link-Grids are calculated for the recently emerging strategies,

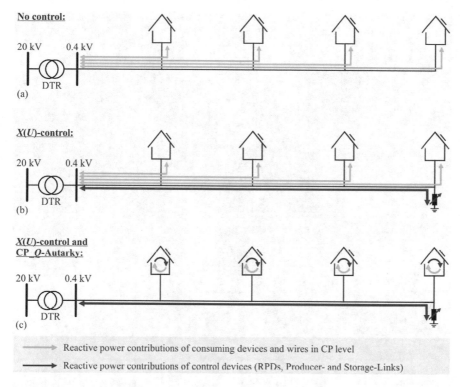

Fig. 4.42 Reactive power flows through an LV feeder for different Volt/var control arrangements: **a** No control; **b** $X(U)$-control; **c** $X(U)$-control and CP_Q-Autarky

i.e. $\cos\varphi(P)$- and $Q(U)$-control of PV inverters and OLTC in distribution substation, and for the newly introduced ones, i.e. $X(U)$-control and its combination with CP_Q-Autarky. The combination of the OLTC in the distribution substation with CP_Q-Autarky is also considered. The focus of this analysis is set on Volt/var controls applied at the LV and CP levels; Volt/var controls of producers connected at the MV level are not considered.

4.7.1 Modelling on the Vertical Axis

The *LINK*-Architecture allows for the systematic analysis of the vertical power system axis [60]. Each Link level may be separately simulated using the extended lumped grid model (see Sect. 4.6.1.3) for the neighbour elements. This approach is illustrated in Fig. 4.43 for the MV, LV, and CP levels. It follows a bottom-up approach with three steps:

1. The first step is to define the CP models, i.e. the structures of the CP_Link-Grids; The $P_t(U_t)$-and $Q_t(U_t)$-behaviour of the connected consuming devices,

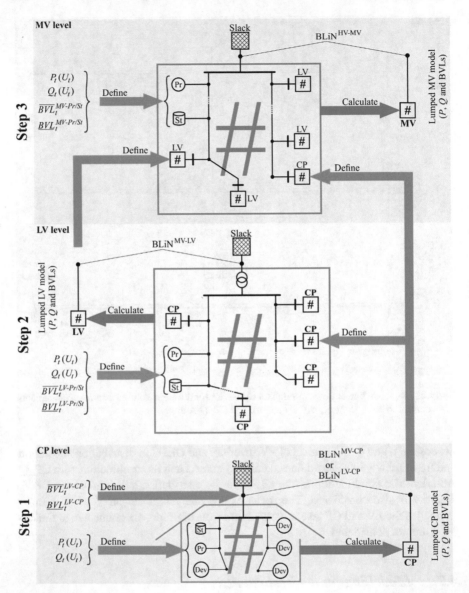

Fig. 4.43 Overview of the systematic modelling of the Link-Grids in the vertical axis

storages and producers; And the upper and lower LV-CP boundary voltage limits. These specifications allow analysing the CP level and calculating the extended lumped CP models according to the procedure described in Sect. 4.6.1.3;

2.   The next step is to analyse the LV level by representing the connected CPs by their extended lumped models and by specifying the behaviour and boundary voltage limits of producers and storages directly connected at the LV level.

Again, the procedure described in Sect. 4.6.1.3 is used to calculate the extended lumped LV_Link-Grid models.

3. Finally, the MV level is analysed by representing the connected LV and CP_Link-Grids by their extended lumped models and by specifying the behaviour and boundary voltage limits of the producers and storages directly connected at the MV level. The extended lumped MV_Link-Grid model can be calculated and provided for the analysis of the HV level.

The daily behaviour of different Link-Grids is calculated for each system level. The following sections Sects. 4.7.2–4.7.4 describe the analysis of selected Link-Grids in detail, while the results of all Link-Grids are catalogued in Sect. A.4.1.

4.7.2 Customer Plant Level

Customer plants comprise the CP_Link-Grid and all thereto connected consuming devices, producers and storages. They are typically unbalanced due to the connection of single-phase devices. The CP_Link-Grid consists of the underpinned wires connecting the house's boundary node to the switches and sockets (see Sect. 2.3.2, Fig. 2.4). The latter ones are the connection points for all consuming devices, electricity producers and storages. CPs are commonly categorised into residential, commercial and industrial ones due to their similar consumption patterns. While residential CPs are always connected at the LV level, commercial and industrial CPs may also be connected at the MV level.

Traditionally, the modelling of single residential CPs was not of particular interest as load flow studies were performed only at the HV and MV level. Lumped models represented the LV grids. These lumped models are usually developed by estimating the average behaviour of many CPs without considering the grid at the LV and CP level [13]. In recent years, the analysis of the LV level became more important due to the integration of distributed generation, electric vehicle chargers, etc. (see Sect. 1.1). Since then, much effort was devoted to develop lumped models of single residential CPs based on load profiles [40, 46] and ZIP models [10, 18]. The load profiles specify the time-dependency of the power contributions at nominal LV-CP boundary voltage. They mainly depend on the behaviour of occupants and thermostatic controls that switch on and off the consuming devices and on the weather conditions that determine the production of PV systems. Meanwhile, the ZIP models specify the voltage-dependency of the power contributions for each instant of time (see Sect. 4.2.2.4). They mainly depend on the load composition, which naturally varies over time.

4.7.2.1 Model Specification

The model of residential CPs located in a rural region is specified for two different types of load profiles: Spiky and smoothed. The spiky one considers the impact of

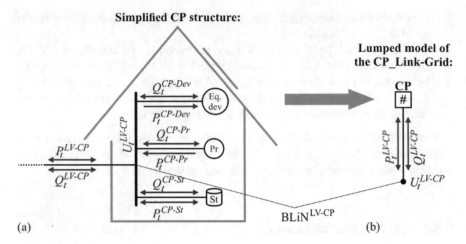

Fig. 4.44 Residential customer plants: **a** Simplified CP structure; **b** Lumped CP_Link-Grid model

clouds on the PV production and reflects the discrete behaviour of the consuming devices and storages within a single CP. Meanwhile, the smoothed one represents the average behaviour of the consuming devices and storages located in many CPs. In both cases, the unbalance and grid at the CP level are neglected, leading to the simplified CP structure shown in Fig. 4.44a. It consists of three components: The equivalent consuming device model (Dev.-model), the producer model (Pr.-model), and the storage model (St.-model), all directly connected to the LV-CP boundary node, $BLiN^{LV-CP}$. The behaviour of the residential CP is analysed for the spiky and smoothed load profiles in Sect. 4.7.2.2.

Model components

The model components are specified to represent rural residential CPs with a PV system, EV charger and modern consuming devices. The spiky load profiles of the Dev.- and St.-model are synthesised with the Load Profile Generator (LPG 2020), which models the behaviour of occupants and thermostatic controls.

Equivalent consuming device model

The Dev.-model represents all consuming devices simultaneously connected to the CP_Link-Grid, including Switch-Mode Power Supply (SMPS), resistive, and lighting devices as well as motors, Eq. (4.40).

$$P_t^{CP-Dev}\left(U_t^{LV-CP}\right) = \sum_{\forall i} P_{i,t}^{CP-Dev}\left(U_t^{LV-CP}\right) \tag{4.40a}$$

$$Q_t^{CP-Dev}\left(U_t^{LV-CP}\right) = \sum_{\forall i} Q_{i,t}^{CP-Dev}\left(U_t^{LV-CP}\right) \tag{4.40b}$$

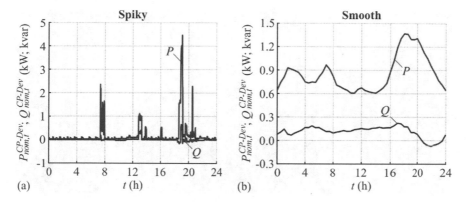

Fig. 4.45 Load profiles of the Dev.-model of the rural residential CP: **a** Spiky; **b** Smoothed

where $P_{i,t}^{CP-Dev}$, $Q_{i,t}^{CP-Dev}$ are the power contributions of the consuming device i. The power contributions of the Dev.-model are specified using load profiles and ZIP coefficients for modern residential CPs, Eq. (4.41).

$$\frac{P_t^{CP-Dev}}{P_{nom,t}^{CP-Dev}} = C^{Z,P} \cdot \left(\frac{U_t^{LV-CP}}{U_{nom}^{LV}}\right)^2 + C^{I,P} \cdot \left(\frac{U_t^{LV-CP}}{U_{nom}^{LV}}\right) + C^{P,P} \tag{4.41a}$$

$$\frac{Q_t^{CP-Dev}}{Q_{nom,t}^{CP-Dev}} = C^{Z,Q} \cdot \left(\frac{U_t^{LV-CP}}{U_{nom}^{LV}}\right)^2 + C^{I,Q} \cdot \left(\frac{U_t^{LV-CP}}{U_{nom}^{LV}}\right) + C^{P,Q} \tag{4.41b}$$

While time-invariant ZIP coefficients from [10] are used for the case with spiky load profiles, time-variant ones from [55] are used for the case with smooth ones. Both the spiky and smoothed load profiles are shown in Fig. 4.45.

The spiky profiles have higher peaks than the smoothed ones but provoke lower daily energy consumptions. While the spiky profiles have consumption peaks of 4.46 kW and 0.65 kvar, the smoothed ones reach maximum consumption values of 1.37 kW and 0.22 kvar. The result is a daily energy consumption of 4.25 and 20.75 kWh, respectively, for the spiky and smoothed load profiles. The maximal reactive power production amounts to 0.16 kvar in Fig. 4.45a and to 0.07 kvar in Fig. 4.45b. The capacitive behaviour in the evening results from the use of LED lamps and other modern appliances.

Producer model

The Pr.-model represents the PV system with a module rating of 5 kW and an inverter rating of 5.56 kVA. Its active power production is determined by the weather conditions and is independent of the LV-CP boundary voltage [74]. Consequently, the production profile alone is sufficient to express the active power part of the Pr.-model, Eq. (4.42). It is characterised by a peak around midday, Fig. 4.46. Clouds reduce the power production, leading to a spiky production profile.

$$P_t^{CP-Pr} = P_{nom,t}^{CP-Pr} \tag{4.42}$$

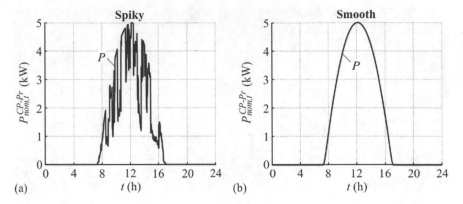

Fig. 4.46 Load profiles of the Pr.-model of the rural residential CP: (a) Time-discontinuous; (b) Time-continuous

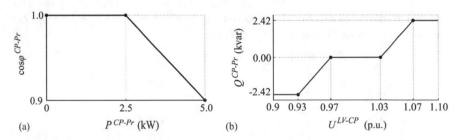

Fig. 4.47 Different control characteristics of the Pr.-model of the rural residential CP: **a** $\cos\varphi(P)$; **b** $Q(U)$

The spiky and smoothed load profiles implicate an energy production of 23.14 and 31.37 kWh, respectively. The reactive power contribution of the Pr.-model is determined by the applied Volt/var control strategy.

Figure 4.47 shows the used $\cos\varphi(P)$- and $Q(U)$-control characteristics. The impact of the $Q(U)$-parametrisation on the behaviour of low voltage grids is discussed in Sect. A.4.2.1. The Pr.-model of a Q-Autarkic CP fully compensates the LV-CP reactive power exchange.

Storage model

The St.-model represents the battery of the EV that is connected through the EV charger. By analogy with the Dev.-model, the active power absorbed by the charger is specified using ZIP models from [64] and load profiles from [6], Eq. (4.43).

$$\frac{P_t^{CP-St}}{P_{nom,t}^{CP-St}} = -0.02 \cdot \left(\frac{U_t^{LV-CP}}{U_{nom}^{LV}}\right)^2 + 0.03 \cdot \left(\frac{U_t^{LV-CP}}{U_{nom}^{LV}}\right) + 0.99 \tag{4.43a}$$

$$Q_t^{CP-St}(U_t) = 0 \tag{4.43b}$$

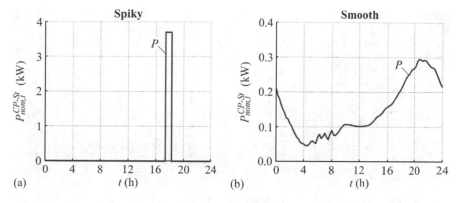

Fig. 4.48 Load profiles of the St.-model of the rural residential CP: **a** Spiky; **b** Smoothed

Figure 4.48 shows the spiky and smoothed load profiles of the St.-model. Residential EV chargers typically draw 3.7 kW at nominal LV-CP boundary voltage, Fig. 4.48a. The charging process of EVs is initiated by the user and terminated when the battery is fully charged or the user unplugs the EV prematurely. Due to the short charging time in the given case, only 1.85 kWh are consumed. The smoothed load profile has a maximum value of 0.30 kW and leads to the daily energy consumption of 3.51 kWh, Fig. 4.48b.

Lumped model equations
The CP_Link-Grid modelling as a single node allows specifying its lumped model, which is shown in Fig. 4.44b, without involving numerical simulations. The corresponding model equations represent the aggregate behaviour of all model components, Eq. (4.44).

$$P_t^{LV-CP} = P_t^{CP-Dev} + P_t^{CP-Pr} + P_t^{CP-St} \tag{4.44a}$$

$$Q_t^{LV-CP} = Q_t^{CP-Dev} + Q_t^{CP-Pr} + Q_t^{CP-St} \tag{4.44b}$$

The LV-CP boundary voltage limits are set to ±10% around the nominal value to reflect the Grid Code requirements [25], Eq. (4.45).

$$\overline{BVL}_t^{LV-CP} = 1.10\,\text{p.u.} \quad \forall t \tag{4.45a}$$

$$\underline{BVL}_t^{LV-CP} = 0.90\,\text{p.u.} \quad \forall t \tag{4.45b}$$

4.7.2.2 Behaviour of CP_Link-Grid

The CP_Link-Grid behaviour is discussed for the case without any Volt/var controlled PV system, for $\cos\varphi(P)$- and $Q(U)$-control, and for CP_Q-Autarky. Both types of load profiles, i.e. spiky and smoothed, are considered separately.

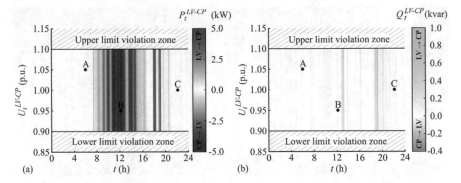

Fig. 4.49 Daily behaviour of the rural residential CP_Link-Grid with spiky load profiles and without any Volt/var control for various voltages at the **a** LV-CP boundary node LV-CP active power exchange;**b** LV-CP reactive power exchange

Spiky load profiles

Without any Volt/var controlled PV system

Constant boundary voltage limits and a strong time-dependency and weak voltage-dependency of the LV-CP active and reactive power exchanges characterise the daily behaviour of the rural residential CP_Link-Grid with spiky load profiles and without any Volt/var control, Fig. 4.49.

The CP mainly consumes active power before 8:00 and after 17:00 and produces it in between. Meanwhile, no clear trend is observed for the reactive power flow direction: It alternates during the day. Before 7:20 and after 21:00, the thermostatic controls that switch on and off the consuming devices dominate the CP power contributions. In cases A and C, the CP consumes 28.60 and 79.17 W and produces 4.37 and 13.07 var, respectively. The active power consumption increases considerably when the occupants actively use consuming devices, i.e. from 7:20 to 8:00 and 17:00 to 21:00. Two active power consumption peaks appear around 18:00 and 19:00. While the former results from EV charging, the latter is provoked by the use of household appliances. Between 8:00 and 17:00, the PV system produces electricity. It turns the consumer into a producer, leading to an active power injection of 4.90 kW and reactive power absorption of 0.02 kvar in case B.

With Volt/var controlled PV system

The Volt/var control strategies do not affect the LV-CP active power exchange because the losses of the PV inverter and the grid at the CP level are not modelled. Therefore, the investigation focus is set on the reactive power exchange. All control strategies significantly modify the customer plant's reactive power behaviour, while the boundary voltage limits are traditionally considered constant, Fig. 4.50.

The $\cos\varphi(P)$-controlled PV system is shown in Fig. 4.50a.

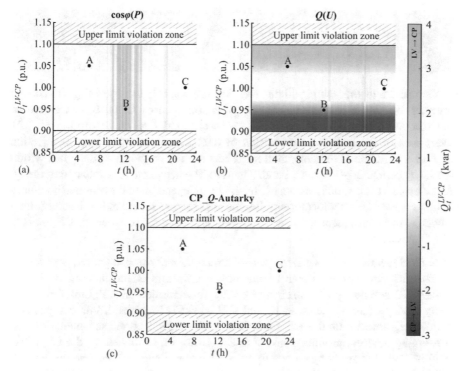

Fig. 4.50 Daily LV-CP reactive power exchange of the rural residential CP_Link-Grid with spiky load profiles for various voltages at the LV-CP boundary node and different control strategies: **a** $\cos\varphi(P)$; **b** $Q(U)$; **c** CP_Q-Autarky

The **$\cos\varphi(P)$control strategy** provokes a strong time-dependency of the reactive power exchange between the LV grid and CPs.

The fluctuating radiation leads to intermittent reactive power contributions. Independently of the boundary voltage, the PV inverter consumes reactive power conform the $\cos\varphi(P)$ characteristic (Fig. 4.47a) when enough active power is produced, i.e. between 9:05 and 14:53. This control strategy drastically increases the total reactive power consumption around noontime. In case B, the reactive power consumption increases from 0.02 kvar when no control is used to 2.44 kvar. Meanwhile, cases A and C remain unaffected.

Figure 4.50b shows the case with $Q(U)$-controlled PV system. According to the $Q(U)$ characteristic (Fig. 4.47b), the PV inverter consumes reactive power for boundary voltages above 1.03 p.u. and produces one for boundary voltages below 0.97 p.u. This control strategy changes the var behaviour of the CP into a mainly voltage-dependent one.

> The **$Q(U)$control strategy** provokes a strong voltage-dependent of the reactive power exchange between the LV grid and CPs.

Compared to the setup without any Vol/var control, the use of $Q(U)$-control reverses the LV-CP reactive power flow direction in cases A and B. In the former, the reactive power injection of 4.37 var is changed into an absorption of 1.21 kvar; And in the latter, the absorption of 0.02 kvar is changed into an injection of 1.19 kvar. Case C remains unaffected because the boundary voltage lies within the $Q(U)$ characteristic's dead-band. The LV-CP reactive power exchange of the Q-Autarkic CP_Link-Grid is shown in Fig. 4.50c. Independent of the time and boundary voltage, the PV inverter locally supplies the reactive power demand of the consuming devices. As a consequence, no reactive power is exchanged between the CP_ and the LV_Link-Grid.

Link-Grid behaviour for different control strategies and the specific cases
The different Volt/var control strategies significantly affect the composition of the LV-CP reactive power exchange of the rural residential CP_Link-Grid with spiky load profiles in cases A, B, and C, Fig. 4.51. In general, this composition includes Q-amount of the CP_Link-Grid itself and all connected elements, i.e. consuming devices, producers and storages. However, in this study, the CP_Link-Grid is modelled as a single node that does not contribute any reactive power. The EV charger is specified to operate with a unity power factor. Therefore, only the consuming devices and producer contribute to the LV-CP reactive power exchange. When spiky load profiles are set, the consuming devices contribute very low amounts of reactive power in the specific cases A, B and C.

When no Volt/var control is applied, the PV system does not contribute any reactive power, so the consuming devices alone determine the LV-CP reactive power exchange. Relatively low Q-amounts result in cases A, B and C. The $\cos\varphi(P)$-controlled PV inverter strongly modifies the reactive power exchange by consuming

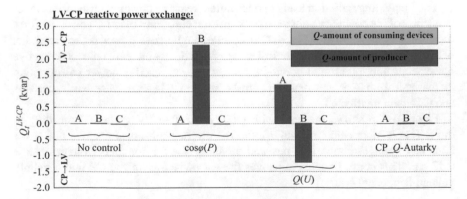

Fig. 4.51 Composition of the LV-CP reactive power exchange of the rural residential CP_Link-Grid with spiky load profiles for different cases, no control and various control strategies

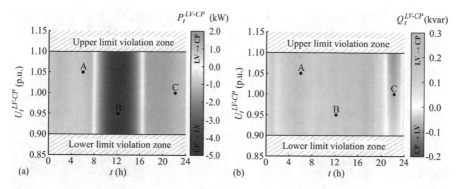

Fig. 4.52 Daily behaviour of the rural residential CP_Link Grid with smoothed load profiles and without any Volt/var control for various voltages at the LV-CP boundary node: **a** LV-CP active power exchange; **b** LV-CP reactive power exchange

2.42 kvar in case B, which is the highest value of all control strategies. Using $Q(U)$-control makes the PV inverter consume and produce 1.21 kvar in cases A and B, respectively. The PV system of a Q-Autarkic CP eliminates the LV-CP reactive power exchange. In this case, the producer mirrors the Q-amount of the consuming devices.

Smoothed load profiles

Without any Volt/var controlled PV system

Figure 4.52 shows the daily behaviour of the rural residential CP_Link-Grid with smoothed load profiles and without any Volt/var control for various voltages at the LV-CP boundary node. As in the case with spiky load profiles, the LV-CP boundary voltage limits are fixed at 0.9 and 1.1 p.u. A significant time-dependency of the power contributions is observed, while the voltage-dependency is rather weak. The smooth character of the load profiles is reflected by smoothly blending colours. Both active and reactive power change their flow directions twice a day.

The CP consumes active psower between 0:00 and 7:40 and between 16:40 and 24:00, reaching values of 0.93 and 1.20 kW for cases A and C, respectively. The uninterrupted radiation leads to an active power production between 8:00 and 16:20. In case B, 4.27 kW are injected. The intervals in which the active power flow changes its direction, i.e. from 7:40 to 8:00 and from 16:20 to 16:40, are characterised by very low active power exchanges. Therein, the consumption is locally supplied by the PV production. The CP consumes reactive power all the time, except for the interval between 21:00 and 23:40, where the LED lamps and other modern appliances shift the inductive CP behaviour into a capacitive one. It consumes 0.18 and 0.14 kvar in cases A and B, respectively, and produces 0.07 kvar in case C.

With Volt/var controlled PV system

Only the reactive power exchange is discussed, as the active power one is not affected by the Volt/var control strategies. Similarly to the case with spiky load profiles, the customer plant's reactive power behaviour is significantly modified by

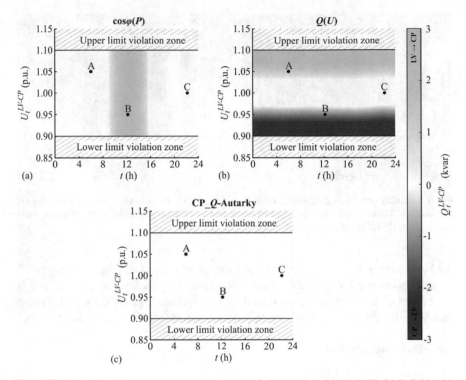

Fig. 4.53 Daily LV-CP reactive power exchange of the rural residential CP_Link-Grid with smoothed load profiles for various voltages at the LV-CP boundary node and different control strategies: **a** $\cos\varphi(P)$; **b** $Q(U)$; **c** CP_Q-Autarky

the control strategies. At the same time, the boundary voltage limits remain unaffected (as they are defined to be 0.9 and 1.1 p.u.), Fig. 4.53.

Fundamentally, the control strategies have the same effects on the LV-CP reactive power exchange as in the case with spiky load profiles: The $\cos\varphi(P)$-control intensifies its time-dependency, while the $Q(U)$-control changes the behaviour into a mainly voltage-dependent one. The former increases the CP's reactive power consumption between 8:53 and 15:28, reaching 2.56 kvar in case B. Meanwhile, the use of $Q(U)$-control provokes an absorption of 1.39 kvar in case A and a production of 0.07 kvar in case C. The CP_Q-Autarky eliminates the LV-CP reactive power exchange.

Link-Grid behaviour for different control strategies and the specific cases

Compared to the CP_Link-Grid with spiky load profiles, the consuming devices consume and produce more reactive power in the selected cases A, B and C, Fig. 4.54. However, the results clearly show the same trend for the control strategies' effect on the LV-CP reactive power exchange.

The PV system without any Volt/var control contributes no reactive power. When $\cos\varphi(P)$-control is applied, it consumes 2.42 kvar in case B. While the $Q(U)$-controlled PV system consumes and produces 1.21 kvar in cases A and B, the one of the Q-Autarkic CP mirrors the reactive power contribution of the consuming devices.

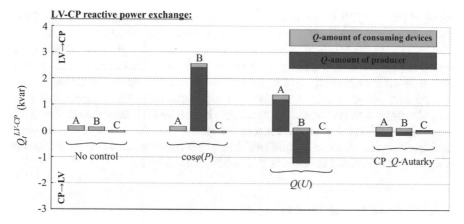

Fig. 4.54 Composition of the LV-CP reactive power exchange of the rural residential CP_Link-Grid with smoothed load profiles for different cases, no control and various control strategies

4.7.3 Low Voltage Level

European low voltage grids are typically unbalanced and radial, and the CPs are connected anywhere along the feeders [12]. On the European average, they have a cable share of 55% [27]. They are roughly categorised into rural and urban ones. While the rural LV_Link-Grids have relative long feeders and low load densities, the urban ones have short feeders and high load densities. Two real Austrian low voltage grids—a rural and an urban one—are analysed for all investigated Volt/var control strategies. Their exact model data is provided in a public data repository [56]. Unbalance is not considered as it is already neglected at the CP level.

4.7.3.1 Model Specification

Figure 4.55 shows the simplified one-line diagram of the rural LV_Link-Grid, wherein the RPDs used for $X(U)$-control and the OLTC are marked as optional elements. It includes four feeders with a total line length of 6.335 km and a cable

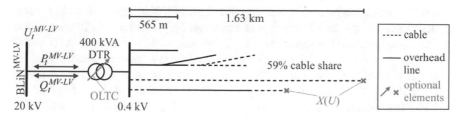

Fig. 4.55 Simplified one-line diagram of the rural LV_Link-Grid

share of 58.64%. While the shortest feeder is 0.565 km in length, the longest one reaches 1.63 km. The feeders connect 61 rural residential CP_Link-Grids.

The 21 kV/0.42 kV distribution transformer is 400 kVA with a total short circuit voltage of 3.7% and a resistive part of 1%. Its OLTC has five tap positions, i.e. 1–5, and adds 2.5% of the nominal LV_Link-Grid voltage per tap. Tap position 3 is the mid position and sets the transmission ratio to its nominal value. When activated, the OLTC maintains the voltage at the secondary bus of the DTR between 0.95 and 0.99 p.u., which is identified as the appropriate setting (see Sect. A.4.2.2). Otherwise, the tap changer is fixed in its mid position. The $X(U)$-controls are parametrized to maintain their terminal voltages between 0.91 and 1.09 p.u.

4.7.3.2 Behaviour of LV_Link-Grid

The behaviour of the LV_Link-Grid is discussed separately for both spiky and smoothed load profiles at the CP level. The MV-LV power exchanges, the LV active power loss, and the DTR loading are analysed in detail. They are calculated without applying any Volt/var control and for the different control strategies.

Spiky load profiles in CP level
While different load profiles are used for all Dev.- and St.-models, the same profile is used for all Pr.-models, reflecting the occupants' individual behaviour and the similar radiation in the region.

Without any Volt/var control strategy
The spikiness of the load profiles at the CP level shapes the daily behaviour of the rural LV_Link-Grid without any Volt/var control, Fig. 4.56. The fluctuating power contributions of the connected CPs deform the upper and lower MV-LV boundary voltage limits into spiky shapes, reaching their minimum and maximum value of 0.9750 and 0.9625 p.u. at 12:23 and 18:44, respectively. No voltage limits are violated in cases A, B and C; But, case $\tilde{B}$ ($U_t^{MV-LV} = 1$p.u., $t = 12:10$) lies within the upper voltage limit violation zone. The power contributions, the active power loss and the DTR loading intensively vary over time. Meanwhile, no distinct dependency on the MV-LV boundary voltage is observable.

Before 7:48 and after 16:32, the LV_Link-Grid absorbs active power from the MV level, while in between, it injects active power. It consumes 8.31 and 28.80 kW in cases A and C, respectively, and produces 257.54 kW in case B. Reactive power is flowing from MV to LV level over the whole day, reaching its highest values during noontime, and the values of 1.18, 23.79 and 3.76 kvar in cases A, B and C, respectively. A considerable active power loss and DTR loading occur during PV production, reaching values of 17.12 kW and 68.06% in case B. In cases A and C, 0.03 and 0.24 kW are lost, respectively, while the DTR is loaded by 2.00 and 7.26%.

The corresponding voltage profiles are shown in Fig. 4.57a, b for cases B and $\tilde{B}$, respectively. While the grey cross marks the MV-LV boundary link node, the black bullets mark the LV-CP ones. Due to the PV injections, the voltages increase with an

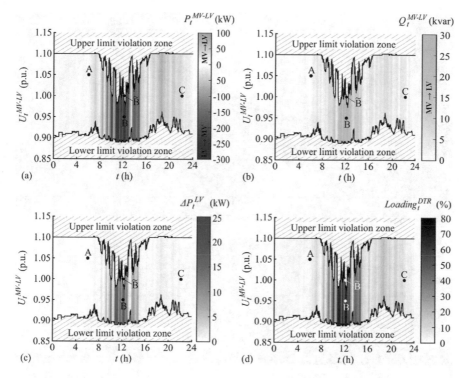

Fig. 4.56 Daily behaviour of the rural LV_Link-Grid without any Volt/var control for various voltages at the MV-LV boundary node and spiky load profiles at the CP level: **a** MV-LV active power exchange; **b** MV-LV reactive power exchange; **c** LV active power loss; **d** DTR loading

increasing distance from the distribution substation in both cases. In Fig. 4.57a, no $BLiN^{LV-CP}$ violates its voltage limits, and therefore, case B lies within the acceptable MV-LV voltage band (see Fig. 4.56). In contrast, Fig. 4.57b shows that violations of the upper voltage limit occur close to the end of the longest feeder in case $\tilde{B}$.

With Volt/var control strategies

Figure 4.58 shows the daily MV-LV reactive power exchange of the rural LV_Link-Grid for various voltages at the MV-LV boundary node, spiky load profiles at the CP level and different control strategies. The combinations of $X(U)$-control and OLTC in the distribution substation with CP_Q-Autarky are denoted as '$X(U) + $' and 'OLTC $+ $'. All Volt/var control strategies strongly modify the MV-LV boundary voltage limits and the reactive power exchange when spiky load profiles are set at the CP level. While the var-based controls, i.e. $\cos\varphi(P)$, $Q(U)$ and $X(U)$, deform the limits, the OLTC shifts them in parallel. All strategies except the $\cos\varphi(P)$ one compress both the upper and lower voltage limit violation zones. The spikiness of the upper BVL^{MV-LV} is effectively mitigated only by the $X(U)$-control and its combination with CP_Q-Autarky. Meanwhile, all var-based controls have a considerable impact on the MV-LV reactive power exchange.

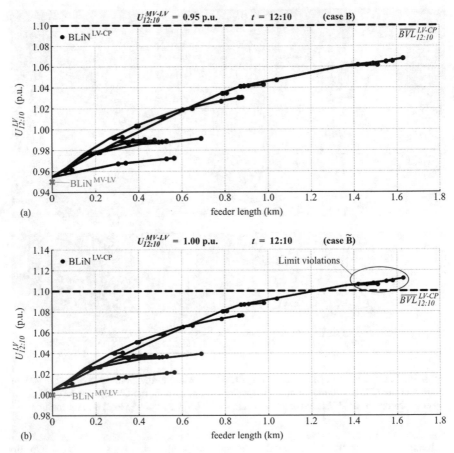

Fig. 4.57 Voltage profiles of the rural LV_Link-Grid's feeders without any Volt/var control at 12:10 for different MV-LV boundary voltages and spiky load profiles at the CP level: **a** 0.95 p.u. (case B); **b** 1.00 p.u. (case B̃)

Figure 4.58a shows the case with $\cos\varphi(P)$-control. The LV_Link-Grid consumes large amounts of reactive power during high PV production periods, tightening the upper limit violation zone significantly and extending the lower one slightly. Although no limit violations occur in case B without any Volt/var control, the $\cos\varphi(P)$-control increases the MV-LV reactive power exchange by 154.51 kvar. In total, 178.29 kvar flow from the MV into the LV level. Meanwhile, cases A and C are not affected by this control strategy. When $Q(U)$-control is applied, the LV grid exchanges large amounts of reactive power in a wide range of the MV-LV boundary voltage, Fig. 4.58b. The additional reactive power flows widen the permissible voltage range over the entire time horizon by compressing both the upper and lower limit violation zones. In cases A and B, the $Q(U)$-control unnecessarily increases the reactive power consumption of the LV-Link-Grid to 37.81 and 36.12 kvar, respectively, while in case C, it does not provoke any additional reactive power flows. The

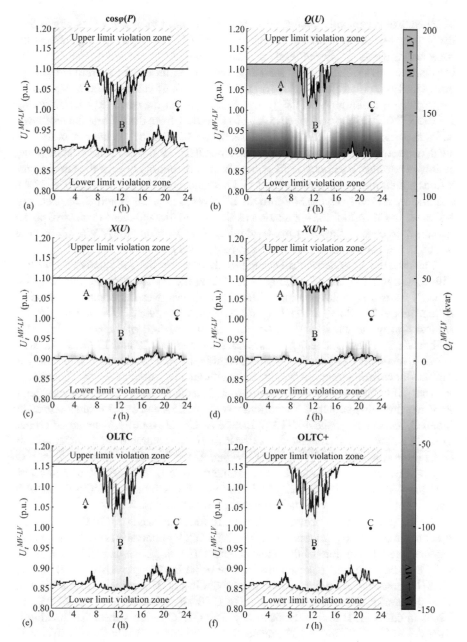

Fig. 4.58 Daily MV-LV reactive power exchange of the rural LV_Link-Grid for various voltages at the MV-LV boundary node, spiky load profiles at the CP level, and different control strategies: **a** $\cos\varphi(P)$; **b** $Q(U)$; **c** $X(U)$; **d** $X(U)$ and CP_Q-Autarky; **e** OLTC; **f** OLTC and CP_Q-Autarky

$X(U)$-control completely changes the shape of the limit violation zones, Fig. 4.58c. Both zones are straightened, allowing for the MV-LV boundary voltage almost the complete range of $\pm$ 10% around nominal voltage during the whole day. The MV-LV reactive power exchange is increased only when required to eliminate voltage limit violations. Therefore, cases A, B and C are not affected by the $X(U)$-control strategy. In combination with CP_Q-Autarky, it reduces the reactive power flow over the BLiN^{MV-LV}, Fig. 4.58d. In case A, 0.16 kvar flow from the LV into the MV level, while in cases B and C, the LV_Link-Grid draws 16.63 and 0.05 kvar from the MV level, respectively. Figure 4.58e shows that the OLTC shifts the limit violation zones mainly in parallel; By about $\pm$ 5%, which is the maximum range of the tap changer. It does not significantly modify the MV-LV reactive power exchange. 1.04, 23.41, and 3.58 kvar flow from the MV into the LV level in cases A, B and C, respectively. As shown in Fig. 4.58f, the application of CP_Q-Autarky reduces the reactive power exchange: In case A, the LV_Link-Grid injects 0.14 kvar into the MV level. In cases B, and C, it absorbs 16.63 and 0.07 kvar, respectively.

The different Volt/var control strategies also modify the daily active power loss within the rural LV_Link-Grid, Fig. 4.59. When the PV systems are injecting, the $\cos\varphi(P)$-control significantly intensifies the grid loss over the complete boundary voltage range: 24.98 kW are lost in case B. Meanwhile, the $Q(U)$-control increases the loss during the whole day but only in the edge regions of the permissible MV-LV boundary voltage range. The losses are increased to 0.29 and 17.68 kW in cases A and B, respectively. The $X(U)$-control significantly increases the grid loss only in regions where limit violations would occur without any Volt/var control. Therefore, it does not affect the losses in the selected cases. The OLTC increases the losses when it decreases the voltages at the LV level and vice versa. 25.4 and 247.5 W are lost in cases A and C, respectively, and 16.31 kW in case B. CP_Q-Autarky, whether combined with $X(U)$-control or OLTC, decreases the losses in the complete voltage–time-plane. In the former, losses of 24.5 W, 17.01 kW and 235.0 W occur in cases A, B, and C, respectively, while in the latter, the losses amount to 25.0 W, 17.01 kW and 242.6 W.

Figure 4.60 shows that the daily DTR loading follows the same trend as the MV-LV reactive power exchange for the different control strategies.

The $\cos\varphi(P)$-control provokes high DTR loadings when the PV systems are injecting. In case B, it amounts to 80.92%. When $Q(U)$-control is used, high loadings appear in the edge regions of the relevant boundary voltage range, reaching 9.23 and 68.32% in cases A and B, respectively. Due to its inactivity in cases A, B, and C, the $X(U)$-control does not affect the corresponding DTR loadings. Its combination with Q-Autarkic CPs slightly unloads the DTR: It is loaded by 1.98, 67.93, and 7.20% in cases A, B and C, respectively. The OLTC reduces the DTR loading in cases A and C to 1.92 and 7.18%, respectively, and increases it to 68.14% in case B. When combined with CP_Q-Autarky, lower DTR loadings occur, i.e. 1.90, 67.93, and 7.13% for cases A, B, and C, respectively.

Link-Grid behaviour for different control strategies and the specific cases

The composition of the MV-LV reactive power exchange differs for the different Volt/var control strategies, Fig. 4.61. It is determined exclusively by the Q-amounts of the connected CP_Link-Grids and the LV_Link-Grid itself, as no storages and

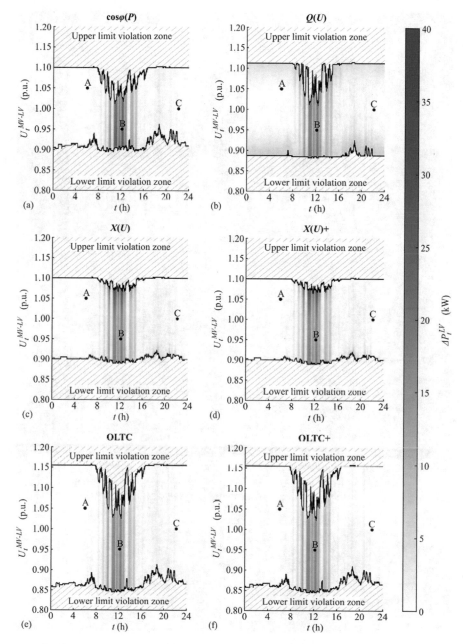

Fig. 4.59 Daily active power loss within the rural LV_Link-Grid for various voltages at the MV-LV boundary node, spiky load profiles at the CP level, and different control strategies: **a** $\cos\varphi(P)$; **b** $Q(U)$; **c** $X(U)$; **d** $X(U)$ and CP_Q-Autarky; **e** OLTC; **f** OLTC and CP_Q-Autarky

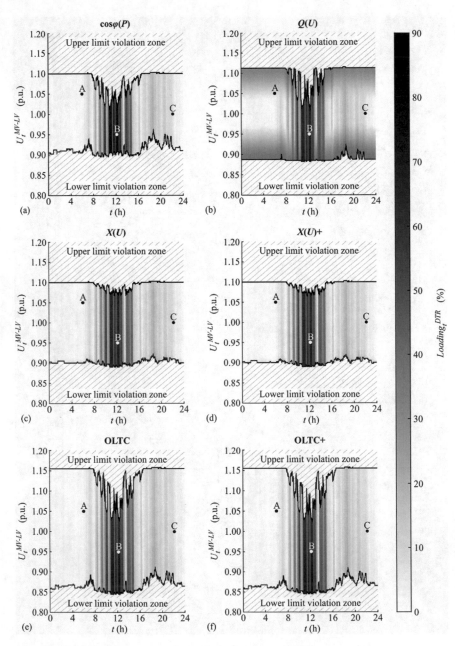

Fig. 4.60 Daily DTR loading within the rural LV_Link-Grid for various voltages at the MV-LV boundary node, spiky load profiles at the CP level, and different control strategies: **a** $\cos\varphi(P)$; **b** $Q(U)$; **c** $X(U)$; **d** $X(U)$ and CP_Q-Autarky; **e** OLTC; **f** OLTC and CP_Q-Autarky

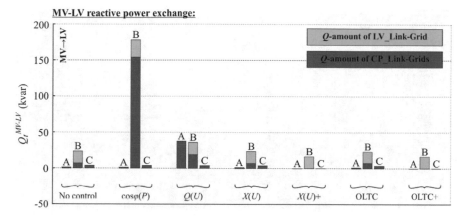

Fig. 4.61 Composition of the MV-LV reactive power exchange of the rural LV_Link-Grid for spiky load profiles at the CP level, different cases, no control and various control strategies

producers are connected at the LV level. Independently of the applied control strategy, the LV lines and the DTR consume significant amounts of reactive power in case B. This reactive power loss results from the intensive active power transfer during PV peak production periods and the low MV-LV boundary voltage (see Eq. 4.4b).

When no control is applied, the Q-amounts of the CP_Link-Grids dominate the reactive power composition in cases A and C, while in case B, the reactive power losses in LV level constitute the major part. In the latter, the use of $\cos\varphi(P)$-control increases the reactive power consumption of the CPs substantially. The additional reactive power flow also increases the Q-losses in LV level considerably. The $Q(U)$-control intensifies the LV-CP reactive power exchanges in cases A and B, which slightly increases the Q-losses at the LV level. The $X(U)$-control is inactive in all cases, yielding the same reactive power composition as the setup without any Volt/var control. Its combination with CP_Q-Autarky eliminates the LV-CP reactive power exchanges and reduces the Q-consumption of the LV Link-Grid marginally. The OLTC has a relatively low impact on the reactive power balance in LV level: In cases A and C, it slightly reduces the Q-amounts of the CPs by reducing the grid voltages. In further consequence, this increases the losses at the LV level. Meanwhile, in case B, the OLTC steps up the voltage, which increases the LV-CP reactive power exchanges and decreases the Q-losses. Its combination with CP_Q-Autarky eliminates the CPs' Q-contributions and the necessity to step up the voltage in case B, reducing the impact of the OLTC on the reactive power losses in all cases.

The distinct Volt/var control strategies differently affect the active power loss within the LV_Link-Grid, Fig. 4.62. In analogy with the reactive power loss, high LV active power losses occur in case B for each control strategy. Meanwhile, relative low losses appear in cases A and C.

Compared to the setup without any Volt/var control, the use of $\cos\varphi(P)$-control considerably intensifies the grid losses in case B. The additional reactive power flows provoked by the $Q(U)$-control strategy slightly increase the active power loss in cases A and B. Due to its inactivity in the selected cases, the $X(U)$-control does not provoke

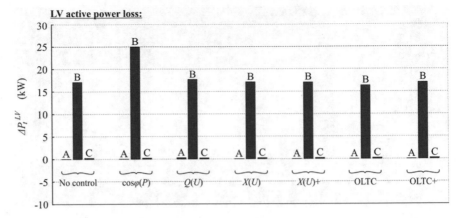

Fig. 4.62 Active power loss within the rural LV_Link-Grid for spiky load profiles at the CP level, different cases, no control and various control strategies

any additional losses. In combination with CP_Q-Autarky, it even reduces the grid loss. The OLTC increases the grid loss in cases A and C and decreases it in case C. This impact is reduced when the OLTC is combined with Q-Autarkic CPs.

The DTR loading is shown in Fig. 4.63 for the different control strategies and cases. The high active power transfer and low MV-LV boundary voltage excessively load the DTR in case B. The $\cos\varphi(P)$-control-related reactive power flows additionally load the DTR in case B. Using $Q(U)$-control increases the DTR loading significantly in case A, and slightly in case B. While the $X(U)$-control does not affect the DTR loading in the selected cases, the OLTC reduces it in cases A and C and increases it in case B. Whether combined with $X(U)$-control or OLTC, the CP_Q-Autarky unloads the DTR in all cases.

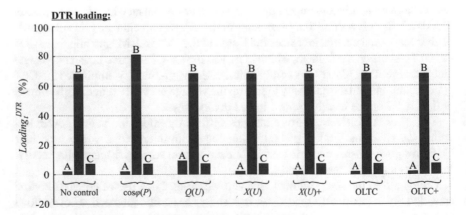

Fig. 4.63 DTR loading within the rural LV_Link-Grid for spiky load profiles at the CP level, different cases, no control and various control strategies

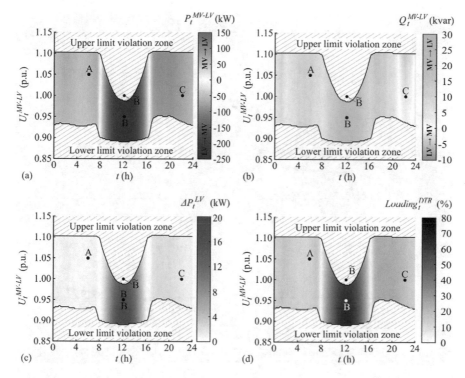

Fig. 4.64 Daily behaviour of the rural LV_Link-Grid without any Volt/var control for various voltages at the MV-LV boundary node and smoothed load profiles at the CP level: **a** MV-LV active power exchange; **b** MV-LV reactive power exchange; **c** LV active power loss; **d** DTR loading

Smoothed load profiles in CP level

The same smoothed load profiles are used for all CPs connected to the rural LV_Link-Grid.

Without any Volt/var control strategy

The smoothed power contributions of the connected CPs deform the boundary voltage limits of the rural LV_Link-Grid into curved shapes and blend the colours representing the power exchanges, the active power loss and the DTR loading continuously, Fig. 4.64. Due to the relatively high power factor at the $BLiN^{MV-LV}$, a strong coupling is observed between the MV-LV active power exchange and the limit curve deformation. In PV production times, the upper BVL is strongly tightened, reaching a very low value of 0.9875 p.u. around noontime. The lower limit is tightened almost during the complete time horizon, especially when the LV_Link-Grid consumes active power. This increases the lower boundary voltage limit up to the maximum value of 0.9525 p.u. in the evening hours. The limits are not violated in cases A, B and C; But, case $\tilde{B}$ lies within the upper limit violation zone. The power exchanges, active power loss and DTR loading are characterised by a strong time-dependency and weak voltage-dependency.

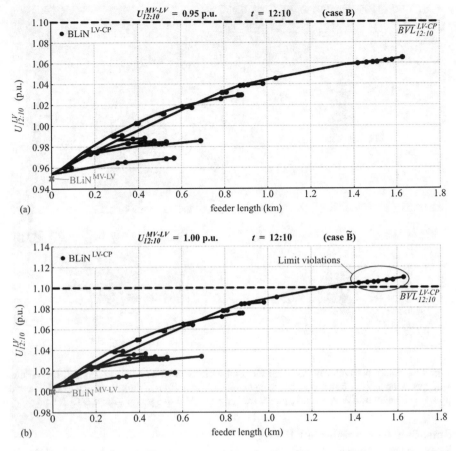

Fig. 4.65 Voltage profiles of the rural LV_Link-Grid's feeders without any Volt/var control at 12:10 for different MV-LV boundary voltages and smoothed load profiles at the CP level: **a** 0.95 p.u. (case B); **b** 1.00 p.u. (case $\tilde{B}$)

The colour shades representing the MV-LV active and reactive power exchanges basically replicate the CP power contributions (see Fig. 4.52). They are only slightly modified by the grid losses, resulting in an MV → LV active power transfer of 56.62 and 73.85 kW in cases A and C, respectively, and a reverse one of 241.99 kW in case B. Meanwhile, 11.21 and 24.78 kvar flow from the MV into the LV level in cases A and B, respectively. In case C, 2.53 kvar are injected into the MV level. The active power loss and the DTR loading are mainly determined by the active power flows, as the reactive power flows are relatively low. While the active power loss amounts to 0.76, 15.97 and 1.39 kW in cases A, B and C, respectively, the DTR is loaded by 13.74, 64.01 and 18.47%.

The corresponding voltage profiles are shown in Fig. 4.65a, b for case B and $\tilde{B}$, respectively. In Fig. 4.65a, no limit violations occur, although the PV injections significantly increase the feeder voltages. Meanwhile, the upper voltage limit is violated in case $\tilde{B}$, Fig. 4.65b.

With Volt/var control strategies

Fundamentally, the same effects of the different Volt/var control strategies on the MV-LV boundary voltage limits and the reactive power exchange are observed for smoothed load profiles in CP level as for the spiky ones, Fig. 4.66. The var-based controls deform the limits, and the OLTC shifts them in parallel. All strategies except the $\cos\varphi(P)$ one compress both the upper and lower voltage limit violation zones. In contrast to the OLTC, the var-based controls significantly affect the MV-LV reactive power exchange.

The $\cos\varphi(P)$-control excessively increases the reactive power flow from the MV into the LV level between 8:53 and 15:28, reaching 179.62 kvar in case B. As a result, the upper boundary voltage limit is significantly tightened, and the lower one is slightly widened. In the remaining interval, neither the BVLs nor the reactive power exchange is affected. In contrast, the $Q(U)$-control compresses both limit violation zones all day through, allowing for boundary voltages above 1.1 and below 0.9 p.u. many hours a day. This considerably intensifies the MV-LV reactive power exchange in the edge regions of the permissible voltage range. Although no limit violations occur without any Volt/var control, the reactive power exchange is increased to 26.02, 35.30 and − 4.08 kvar in cases A, B and C, respectively. $X(U)$-control, whether combined with CP_Q-Autarky or not, straightens the upper and lower BVLs, leaving only small protrusions during the noon and evening hours. It is inactive in the selected cases, as no reactive power support is necessary to maintain acceptable voltages. Therefore, when $X(U)$-control is combined with Q-Autarkic CPs, the MV-LV reactive power exchange is reduced to 0.52, 15.19 and 1.15 kvar in cases A, B and C, respectively. The OLTC shifts the BVLs by around $\pm$ 5% in parallel, conserving their original shape. This slightly reduces the MV-LV reactive power exchange to 10.46 and −2.06 kvar in cases A and C, respectively. It is further reduced when the CPs act Q-Autarkic: 0.54, 15.19 and 1.18 kvar flow from the MV level into the LV level.

For smoothed load profiles at the CP level, similar effects of the different control strategies on the LV active power loss are observed as for the spiky ones, Fig. 4.67. The $\cos\varphi(P)$-control drastically increases the grid loss around noontime, reaching 23.92 kW in case B. $Q(U)$-control adds significant losses only in the edge regions of the permissible voltage range, provoking values of 0.83, 16.44 and 1.40 kW in cases A, B and C, respectively. The $X(U)$-control is active mainly in regions where limit violations would occur without any Volt/var control. Therefore, it does not affect the losses in the selected cases. Meanwhile, its combination with CP_Q-Autarky reduces the losses to 0.74, 15.86 and 1.39 kW in cases A, B and C, respectively. The OLTC steps down the voltage in cases A and C, reducing the loss to 0.76 kW in case A and increasing it to 1.418 kW in case C. The losses resulting from this setup are reduced to 0.74, 15.86 and 1.416 kW when the CPs act Q-Autarkic.

Also, for the DTR loading, analogous effects for smoothed and spiky load profiles at the CP level are observed, Fig. 4.68.

The use of $\cos\varphi(P)$-control highly loads the DTR when the PV systems produce significant amounts of active power, reaching 77.96% in case B. The additional reactive power flows provoked by the $Q(U)$-control increase the DTR loading in the edge region of the permissible voltage range. In cases A, B and C, this loads the

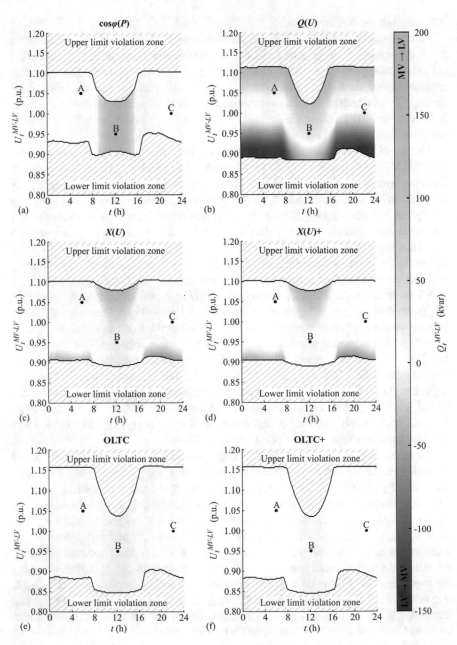

Fig. 4.66 Daily MV-LV reactive power exchange of the rural LV_Link-Grid for various voltages at the MV-LV boundary node, smoothed load profiles at the CP level and different control strategies: **a** $\cos\varphi(P)$; **b** $Q(U)$; **c** $X(U)$; **d** $X(U)$ and CP_Q-Autarky; **e** OLTC; **f** OLTC and CP_Q-Autarky

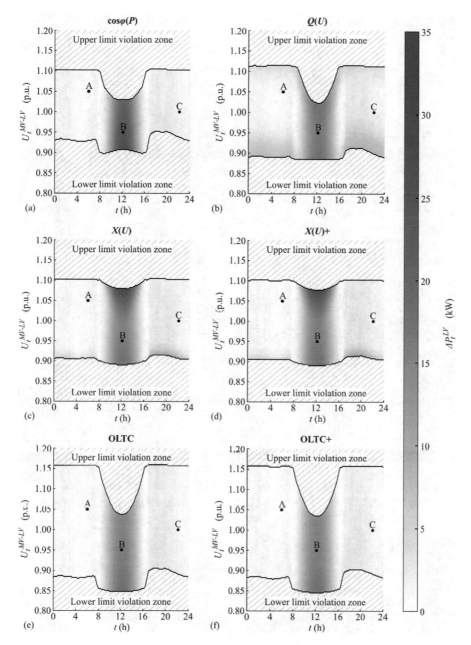

Fig. 4.67 Daily active power loss within the rural LV_Link-Grid for various voltages at the MV-LV boundary node, smoothed load profiles at the CP level and different control strategies: **a** $\cos\varphi(P)$; **b** $Q(U)$; **c** $X(U)$; **d** $X(U)$ and CP_Q-Autarky; **e** OLTC; **f** OLTC and CP_Q-Autarky

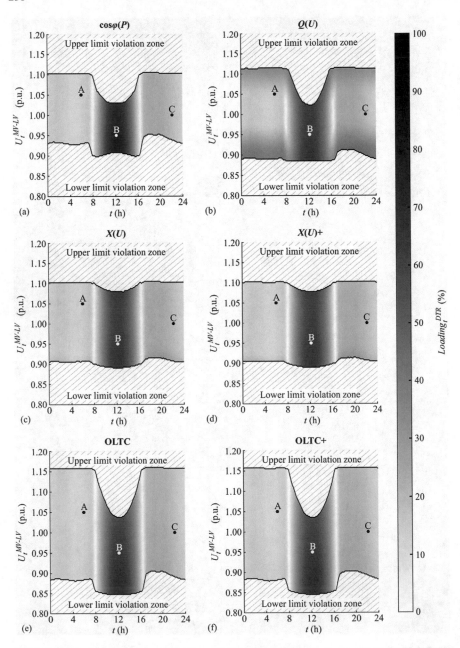

Fig. 4.68 Daily DTR loading within the rural LV_Link-Grid for various voltages at the MV-LV boundary node, smoothed load profiles at the CP level, and different control strategies: **a** $\cos\varphi(P)$; **b** $Q(U)$; **c** $X(U)$; **d** $X(U)$ and CP_Q-Autarky; **e** OLTC; **f** OLTC and CP_Q-Autarky

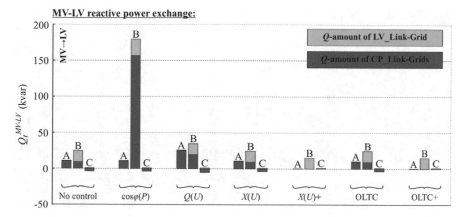

Fig. 4.69 Composition of the MV-LV reactive power exchange of the rural LV_Link-Grid for smoothed load profiles at the CP level, different cases, no control and various control strategies

DTR by 14.82, 64.27 and 18.50%, respectively. As not active in these cases, the $X(U)$-control does not affect the corresponding DTR loadings. Its combination with Q-Autarkic CPs slightly unloads the DTR to 13.51, 63.81 and 18.46% in cases A, B and C, respectively. The DTR loading is decreased in cases A and C when the OLTC is used to control the voltage, obtaining values of 13.05 and 18.21%, respectively. Its combination with CP_Q-Autarky further reduces it to 12.84, 63.81 and 18.20% respective to cases A, B and C.

Link-Grid behaviour for different control strategies and the specific cases

Figure 4.69 shows the composition of the reactive power exchanged between MV_ and LV_Link-Grid. Significant reactive power losses occur in case B for each control strategy.

Without any Volt/var control, the Q-composition is dominated by the Q-amounts of the CPs in cases A and C and by the grid loss in case B. In the latter, $\cos\varphi(P)$-controlled PV systems substantially increase the LV-CP reactive power exchanges, causing additional reactive power losses in further consequence. The $Q(U)$-control intensifies the LV reactive power loss and especially the LV-CP reactive power exchanges in all cases. Due to its inactivity in the selected cases, the $X(U)$-control does not modify the corresponding reactive power compositions. In cases A and C, the OLTC reduces the reactive power Q-amounts of the CPs and increases the one of the LV_Link-Grid itself. Q-Autarkic customers do not exchange any reactive power with the grid, reducing the grid's reactive power loss in both combinations.

When smoothed load profiles are used at the CP level, the different Volt/var control strategies affect the composition of the MV-LV reactive power exchange in the same way as when spiky load profiles are used.

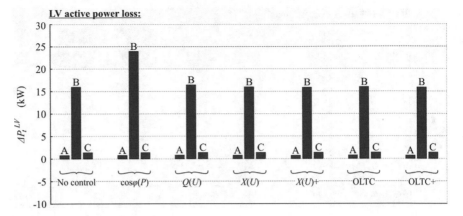

Fig. 4.70 Active power loss within the rural LV_Link-Grid for smoothed load profiles at the CP level, different cases, no control and various control strategies

The active power loss in LV level behaves analogue to reactive power one, Fig. 4.70. While high losses occur in case B for each control strategy, relative low ones prevail in cases A and C.

While the $\cos\varphi(P)$-control considerably intensifies the grid losses in case B, the $Q(U)$-control slightly increases the losses in all cases. The $X(U)$-control does not affect the losses in the selected cases. When an OLTC is used, the losses are decreased in case A and increased in case C. The application of CP_Q-Autarky generally reduces the grid loss.

The impact of the different Volt/var control strategies on the DTR loading follows the same trend for smoothed as for spiky load profiles at the CP level. However, significantly higher values are observed for the former in cases A and C, Fig. 4.71.

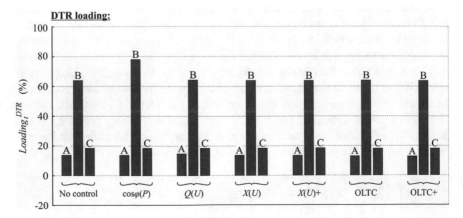

Fig. 4.71 DTR loading within the rural LV_Link-Grid for smoothed load profiles at the CP level, different cases, no control and various control strategies

Compared to the $\cos\varphi(P)$-control, which drastically increases the DTR loading in case B, the other control strategies have a marginal impact. In contrast to the $X(U)$-control, the $Q(U)$-control slightly increase the DTR loading in all cases. Using an OLTC reduces the DTR loading in cases A and C. In any combination, Q-Autarkic CPs unload the DTR from their reactive power contributions.

4.7.4 Medium Voltage Level

European medium voltage grids are three-phase and basically balanced. They are of meshed or radial structure, with the latter dominating rural installations. Each feeder includes numerous laterals with connected distribution substations [12]. On the European average, they have a cable share of 41% [27]. Two real Austrian medium voltage grids—a large and a small one—are analysed for all investigated Volt/var control strategies.

4.7.4.1 Model Specification

Figure 4.72 shows a large MV_Link-Grid, which includes six feeders with a total line length of 267.151 km and a cable share of 74.66%. While the shortest feeder is 2 km in length, the longest one reaches 46.10 km. It is operated with a nominal voltage of 20 kV. In total, the feeders connect 143 commercial and two big industrial CP_Link-Grids, as well as 45 rural and 11 urban LV_Link-Grids. Furthermore, 15 hydroelectric power plants with maximal production capacities between 60 and 400 kW are connected along the feeders. They are simply modelled as PQ node-elements that constantly inject 70% of their maximal active power production with unity power factor. Their upper and lower BVLs are set to 1.1 and 0.9 p.u., respectively. The supplying transformer is not included in the model.

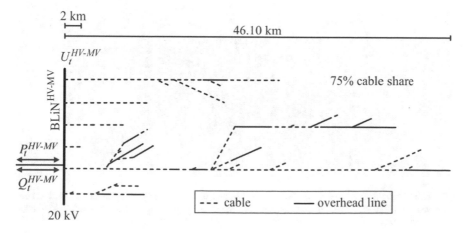

Fig. 4.72 Simplified one-line diagram of the large MV_Link-Grid

4.7.4.2 Behaviour of MV_Link-Grid

The analysis of the CP and LV level has revealed the same trends in the Volt/var behaviour for both types of load profiles: Spiky and smoothed. Therefore, the MV level is calculated only for the smoothed load profiles at the CP level. The HV-MV power exchanges and the MV active power loss are analysed in detail. They are calculated without applying any Volt/var control and for the different control strategies.

Without any Volt/var control strategy

   The boundary voltage limits at the BLiN^{HV-MV} of the large MV_Link-Grid are drastically deformed, leaving only a small corridor of permissible voltages during the day, Fig. 4.73. Each point within the limit violation zones corresponds to violations of the connected elements' BVLs, i.e. commercial and industrial CPs directly connected to the MV_Link-Grid, hydroelectric power plants, and urban and rural LV_Link-Grids. The colours representing the HV-MV power exchanges and the active power loss blend into each other continuously. Strong coupling is found between the active power exchange and the limit curve deformation. The injection of active power into the HV level lowers the upper BVL to 0.9175 p.u. at midday. As a consequence,

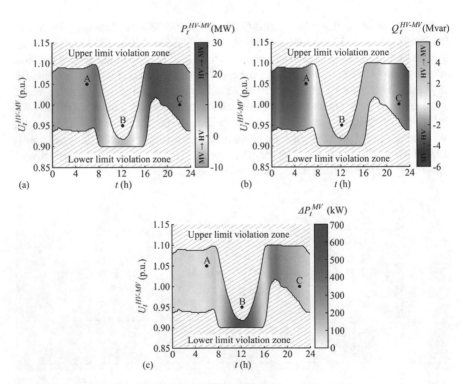

Fig. 4.73 Daily behaviour of the large MV_Link-Grid without any Volt/var control for various voltages at the HV-MV boundary node and smoothed load profiles at the CP level: **a** HV-MV active power exchange; **b** HV-MV reactive power exchange; **c** MV active power loss

case B lies far within the upper limit violation zone. When high amounts of active power flow into the MV level, the lower voltage limit increases, reaching 1.0175 p.u. at 18:00. A strong time-dependency and weak voltage-dependency are observed for the power exchanges and the active power loss. The results clearly show that MV_Link-Grids with high PV penetration can hardly be operated in the traditional way, i.e. with OLTC in supplying substation and without additional Volt/var control strategies applied in LV and/or CP level.

The MV_Link-Grid injects active power into the HV level from 10:15 to 13:59 and absorbs one before and after this interval. In case A and C, 12.22 and 15.21 MW flow from the HV into the MV level, respectively, while in case B, 6.22 MW flow reversely. Meanwhile, the reactive power flows into the HV level at night-time, i.e. before 7:19 and after 20:12, amounting to 2.18 and 2.29 Mvar respective to cases A and C. In case B, the MV_Link-Grid absorbs 5.32 Mvar from the HV level. Relative high active power losses occur around midday and around 18:00. 83.11, 536.98 and 144.18 kW are lost in cases A, B and C, respectively.

Figure 4.74 shows the corresponding voltage profiles for case B. Therein, the boundary nodes to the connected elements are marked by different symbols and colours: While black bullets are used to represent the boundary link nodes to the commercial and industrial CPs, the ones to the rural and urban LV_Link-Grids are marked by yellow and violet asterisks, respectively. The boundary producer nodes at which the hydroelectric power plants are connected are highlighted as red asterisks. Furthermore, the relevant boundary voltage limits are shown in the respective colours. The CPs and the hydroelectric power plants all have an upper boundary voltage limit of 1.1 p.u. throughout the complete time horizon.

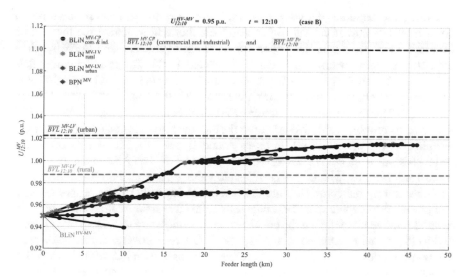

Fig. 4.74 Voltage profiles of the large MV_Link-Grid's feeders without any Volt/var control at 12:10 for an HV-MV boundary voltage of 0.95 p.u. (case B) and smoothed load profiles at the CP level

Meanwhile, the BVLs of the rural and urban LV_Link-Grids vary over time, restricting the maximal MV-LV boundary voltage to 0.9875 and 1.0225 p.u. at 12:10, respectively (see Figs. 4.125 and 4.134 in Sect. A.4.1.2). The voltages increase along the feeders, reaching 1.0161 p.u. close to the end of the longest feeder. In these conditions, some of the BLiN^{MV-LV} to the rural LV_Link-Grid violate their upper voltage limit. As a result, case B lies within the upper limit violation zone.

Figure 4.74 also shows that no LV_Link-Grids are connected to two relative short MV feeders. This should be kept in mind when comparing the different Volt/var control strategies: In contrast to the PV inverter based controls, the $X(U)$-control and the OLTC in distribution substation do not affect the voltage profiles of these MV feeders.

With Volt/var control strategies

The investigated Volt/var control strategies differently widen the permissible voltage band at the HV-MV boundary link node and differently affect the corresponding reactive power exchange, Fig. 4.75. Each control strategy eliminates the limit violations in case B. As observed at the LV level, the var-based controls, i.e. $\cos\varphi(P)$, $Q(U)$ and $X(U)$, deform the limits also at the MV level. Meanwhile, the parallel shifting effect of the OLTC is restricted by the fact that the commercial and industrial CPs and the hydroelectric power plants do not include transformers with OLTCs.

All strategies except the $\cos\varphi(P)$ control compress both the upper and lower voltage limit violation zones. Only the var-based controls significantly modify the HV-MV reactive power exchange.

During the daytime, the $\cos\varphi(P)$-control provokes tremendous reactive power flows from HV into MV level, tightening the upper and widening the lower limit violation zone. It relaxes the upper voltage limit at midday to 1.0025 but does not ease the highly restricting lower BVL around 18:00. The reactive power flow amounts to 21.19 Mvar in case B. $Q(U)$-control affects the var exchange almost in the complete voltage–time-plane, compressing both the upper and lower violation zones significantly. At midday, the upper HV-MV boundary voltage limit of 0.965 p.u. remains relative restrictive. Meanwhile, the lower limit is considerably relaxed, reaching 0.9425 p.u. at 18:00. In cases A and B, the MV_Link-Grid draws 0.11 and 6.40 Mvar from the HV level, while in case C, it injects 2.59 Mvar. The $X(U)$-control significantly widens the permissible voltage band at the HV-LV boundary link node during the complete time horizon by adding only small portions of reactive power. While the upper BVL is greatly relaxed around midday, the lower one remains relative restrictive in the evening hours: Voltages up to 1.035 p.u. and down to 0.9775 p.u. are acceptable at midday and 18:00, respectively. The HV-MV reactive power exchange is insignificantly affected in cases A and C, while in case B, it is increased to 5.52 Mvar. The OLTC increases the upper and decreases the lower BVLs during the complete time horizon without changing the var behaviour considerably. However, both limits remain relative restrictive: Maximal and minimal HV-MV boundary voltages of 0.9725 p.u. are allowed at midday and 18:00, respectively.

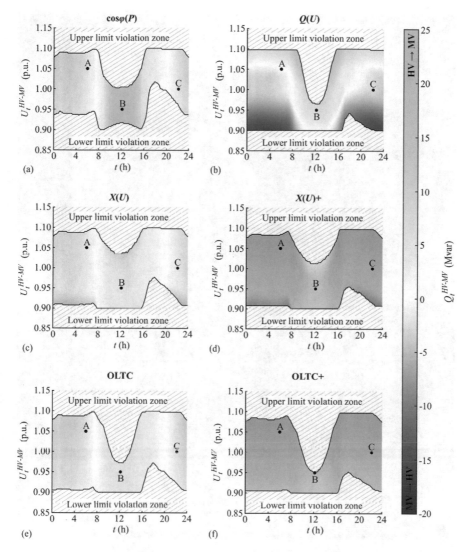

Fig. 4.75 Daily HV-MV reactive power exchange of the large MV_Link-Grid for various voltages at the HV-MV boundary node, smoothed load profiles at the CP level and different control strategies: **a** $\cos\varphi(P)$; **b** $Q(U)$; **c** $X(U)$; **d** $X(U)$ and CP_Q-Autarky; **e** OLTC; **f** OLTC and CP_Q-Autarky

In cases A and C, 2.23 and 2.27 Mvar flow from the MV into the HV level. Meanwhile, 5.33 Mvar flow reversely in case B. Whether combined with $X(U)$-control or OLTC, the CP_Q-Autarky reduces the upper and lower BVLs. It intensifies the capacitive behaviour seen from the HV level in cases A and C and reverses the reactive power flow in case B. In the former combination, the MV_Link-Grid injects 7.48, 4.75 and 6.58 Mvar, respectively, while in the latter, it injects 7.48, 5.14 and 6.58 Mvar.

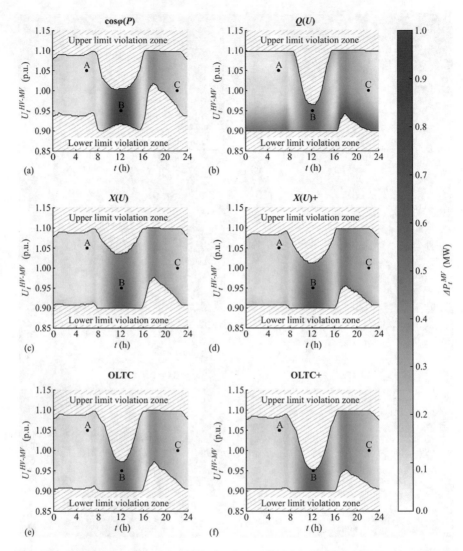

Fig. 4.76 Daily active power loss within the large MV_Link-Grid for various voltages at the HV-MV boundary node, smoothed load profiles at the CP level and different control strategies: **a** $\cos\varphi(P)$; **b** $Q(U)$; **c** $X(U)$; **d** $X(U)$ and CP_Q-Autarky; **e** OLTC; **f** OLTC and CP_Q-Autarky

The Volt/var control strategies differ from each other in the active power loss they provoke in MV level, Fig. 4.76.

Intensive losses result from the use of the $\cos\varphi(P)$-control around noontime, reaching 846.09 kW in case B. The $Q(U)$-control increases the active power loss for boundary voltages close to the lower limit and decreases it for boundary voltages close to the upper limit. This reduces the loss to 52.61 kW in case A and increases it to 538.54 and 149.24 kW cases B and C, respectively. While the $X(U)$-control does not significantly affect the losses in MV level, its combination with Q-Autarkic CPs increases them almost in the complete voltage–time plane. Losses of 121.45, 554.80 and 181.14 kW occur in cases A, B and C, respectively. The OLTC has a very low impact on the MV active power loss, provoking 81.92, 537.24 and 143.63 kW respective to cases A, B and C. When combined with CP_Q-Autarky, higher losses of 119.65, 572.32 and 180.34 kW occur in these cases.

Link-Grid behaviour for different control strategies and the specific cases

The MV_Link-Grid itself and the connected CP_ and LV_Link-Grids contribute reactive power, thus composing the HV-MV reactive power exchange. Meanwhile, the hydroelectric power plants do not produce and consume any reactive power. The compositions resulting in cases A, B and C strongly depend on the applied control strategy, Fig. 4.77. Due to its high cable share, the MV_Link-Grid generally produces significant amounts of reactive power. Especially in case B, this reactive power production is partly compensated by the reactive power losses in the MV lines' series impedances.

When no Volt/var control is applied, the connected CPs draw reactive power from the MV level in any case. The LV_Link-Grids absorb relative low amounts of reactive power in cases A and B, while in case C, they behave slightly capacitive and inject reactive power. The reactive power consumption is overcompensated in cases A and C, leading to a capacitive behaviour of the whole distribution grid at the HV-MV boundary node. This behaviour is changed into an inductive one by the large consumption of the CPs and LV_Link-Grids in case B. The use of cos$\varphi(P)$-control drastically increases the reactive power contributions of the CP_ and especially the LV_Link-Grids in case B. Tremendous amounts of reactive power are absorbed from the HV level. $Q(U)$-controlled PV inverters increase the reactive power consumption of the CPs in case A and decrease it in case B; No significant modification occurs in case C. Furthermore, they intensify the reactive power contribution of the LV_Link-Grids in all cases. The $X(U)$-control does not significantly affect the reactive power

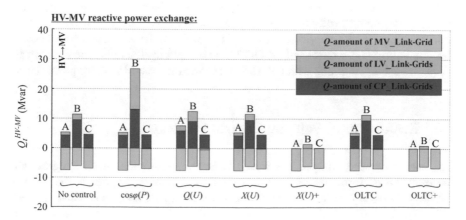

Fig. 4.77 Composition of the HV-MV reactive power exchange of the large MV_Link-Grid for smoothed load profiles at the CP level, different cases, no control and various control strategies

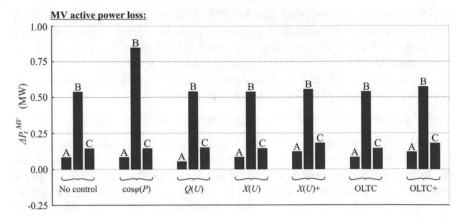

Fig. 4.78 Active power loss within the large MV_Link-Grid for smoothed load profiles at the CP level, different cases, no control and various control strategies

composition in cases A and C, while in case B, it slightly increases the reactive power absorption of LV_Link-Grids. This reduces the voltages in MV level and thus the reactive power consumption of the connected CPs. Meanwhile, the OLTC in the distribution substation has a very low impact on the Q-composition at the MV level. In both combinations, the Q-Autarky eliminates the relative large reactive power consumption of the commercial and industrial CPs, drastically changing the overall system behaviour. It also reduces the reactive power contributions of the LV_Link-Grids in all cases, reversing MV-LV reactive power exchange in case C in total.

The different Volt/var control strategies provoke different active power losses in MV level, Fig. 4.78. Due to the intensive power transfer, relatively high losses prevail in case B, while lower ones occur in case C and especially in case A.

The $\cos\varphi(P)$-control considerably intensifies the loss in case B, as it drastically increases the reactive power flows through the MV lines. Meanwhile, the $Q(U)$-control partly compensates the reactive power production of the MV lines in case A, thus reducing the reactive power flows in MV level and the associated active power loss. Both the $X(U)$-control and the OLTC do not significantly affect the losses in all cases. Their combinations with CP_Q-Autarky increase the MV active power loss.

4.8   Evaluation of Volt/var Control Strategies

The comprehensive analysis conducted in Sect. 4.7 provides deep insights into the technical performance of the Volt/var control strategies applied at the LV and CP level. Volt/var control of producers connected in MV level is not investigated. However, the large number of simulation results makes it hard to recognise each control strategy's strengths and weaknesses at a glance. Social aspects, such as data privacy and discrimination, are not considered.

This section provides a clear comparison of the Volt/var control strategies by evaluating them against various technical and social criteria. The evaluation results are visualised for both LV_Link-Grids, which are catalogued in Sect. A.4.1.2, within separate evaluation hexagons.

4.8.1 Evaluation Procedure

The compact comparison of the control strategies requires the specification of the relevant criteria and their visualisation within the hexagon. Both are roughly described in Sects. 4.8.1.1 and 4.8.1.2. The exact calculation formulas are given in Sect. A.4.3.

4.8.1.1 Evaluation Criteria

The performance of the different control strategies is assessed by means of technical and social criteria. While the technical ones are calculated based on the simulation results, the social ones are analysed qualitatively.

- Technical criteria

    - **Voltage limit violations**: The purpose of the investigated Volt/var control strategies is to mitigate violations of the stipulated voltage limits of ±10 % around the nominal value. This impact is assessed using the voltage limit Violation Index (VI), which penalises limit violations at the LV level by considering the number of violating nodes and the corresponding voltage values.
    - **MV-LV reactive power exchange**: The impact of the control strategies on the reactive power flow through the MV-LV boundary node is assessed by means of the reactive energy exchange. The flow direction is not considered for the calculation.
    - **Active power loss**: Each control strategy provokes distinct active power losses within the grid. This impact is evaluated based on the energy loss at the LV level.
    - **DTR loading**: The control strategies modify the DTR loading. This impact is assessed by calculating the average loading of the DTR.

The technical criteria are calculated for the (U,t)-plane spanned by the simulated time horizon of 24 hours and by the MV-LV boundary voltages between 0.9 and 1.1 p.u. This plane is shown in Figure 4.79 and is denoted as 'Evaluation zone'.

Fig. 4.79 Zone within the (U,t)-plane used to calculate the technical evaluation criteria for all simulations in LV level

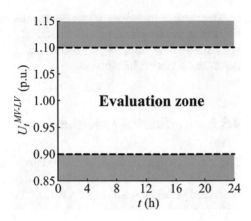

- Social criteria

  - **Discrimination**: Per definition, all individuals within a non-discriminatory society have an equal and fair prospect to access the available opportunities. In power system operation, an opportunity for customers, storage- and producer-operators is to provide reactive power as an ancillary service to support the grid operation. To promote freedom from discrimination in power system operation, [22] directs DSOs to procure the non-frequency ancillary services needed for their systems according to transparent, non-discriminatory, and market-based procedures. In this sense, the applied control strategy should enable all stakeholders an equal and fair prospect to offer reactive power. The discrimination criterion is set to one when the reactive power contribution duty is divided unequally between the customers, and otherwise, it is set to zero.
  - **Data privacy**: The privacy of each individual, customer and company is inviolable. The applied control strategy should enable the coordination of the underlying control variables with minimal data exchanges through external interfaces to protect data privacy. The data privacy criterion is set to one when data exchanges through the LV-CP interfaces are necessary to coordinate the control variables, and otherwise, it is set to zero.

4.8.1.2   Result Visualisation

The visualisation within the evaluation hexagon requires further processing of the calculated technical criteria (see A.4.3.2). The calculated values of the evaluation criteria are normalised to eliminate their physical units and lie within the interval [0, 1].

Figure 4.80 shows the evaluation hexagon with the worst and ideal performance of the control strategies. On its corners are set the evaluation criteria. A completely filled area indicates the worst performance, and the ideal one corresponds to a point in the middle of the chart. Violations of the voltage limits offend against the law. Thus, they are unacceptable. The corresponding zone within the evaluation hexagon is highlighted in red.

Fig. 4.80 Evaluation
hexagon with the worst and
ideal performance of the
control strategies

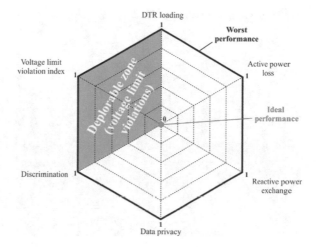

4.8.2   Evaluation Results

The voltage limit violation index is the most critical technical evaluation criterion.
It is shown in Fig. 4.81 for the rural and urban LV_Link-Grids and the investigated
control strategies. The violation index is always higher in the rural LV_Link-Grid
than in the urban one.

A clear trend is observed in both Link-Grids. All control strategies mitigate the
excessive voltage limit violations that occur without any Volt/var control. However,
relative high violation indices remain when $\cos\varphi(P)$-control is used, which is mainly
caused by its inability to mitigate the lower voltage limit violations. In comparison, lower but still significant violation indices result from $Q(U)$-control. The
OLTC provokes relative low voltage limit violation indices compared to the $\cos\varphi(P)$-

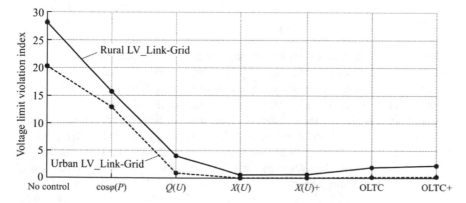

Fig. 4.81 Voltage limit violation index for the rural and urban LV_Link-Grids and different control
strategies

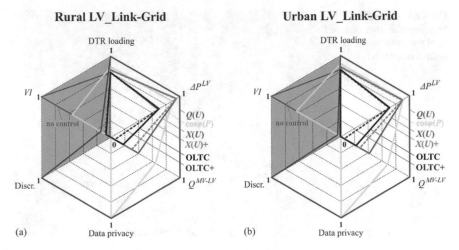

Fig. 4.82 Evaluation hexagons of Volt/var control strategies for different LV_Link-Grids: **a** Rural; **b** Urban

and $Q(U)$-control, especially within the urban LV_Link-Grid. The lowest violation indices are reached in both LV_Link-Grids when $X(U)$-control is used. Combining the OLTC and the $X(U)$-control with the CP_Q-Autarky slightly increases the corresponding violation indices.

Figure 4.82a, b show the evaluation hexagons for the rural and urban LV_Link-Grids separately. Different colours and line types present the control strategies: Solid lines in yellow, orange, green, purple and grey are used for $\cos\varphi(P)$, $Q(U)$, $X(U)$, OLTC in distribution substation, and for the setup without any Volt/var control, respectively. The combinations of $X(U)$ and OLTC with Q-Autarkic CPs are shown by dashed lines in lighter shades of the corresponding colours. They are designated as '$X(U)$ + ' and 'OLTC + ', respectively.

While clear trends are observed in both LV_Link-Grids for the voltage limit violation index, the DTR loading and the MV-LV reactive power exchange, none is found for the active power loss. $\cos\varphi(P)$-control provokes an unacceptably high voltage limit violation index. Furthermore, a relative high DTR loading, active power loss, and MV-LV reactive power exchange appear. It does not guarantee data privacy but avoids discriminatory ancillary service procurement. The highest DTR loading, active power loss and MV-LV reactive power exchange occur when $Q(U)$-control is used. It leaves significant violations of the voltage limits, and neither supports data privacy nor non-discriminatory ancillary service procurement. $X(U)$-control eliminates almost all voltage limit violations while increasing the DTR loading and MV-LV reactive power exchange to a moderate extent. It provokes a relative high active power loss while preserving data privacy and freedom of discrimination. Its combination with CP_Q-Autarky significantly reduces the MV-LV reactive power exchange. The active power loss and DTR loading are slightly reduced, and a few more violations of the voltage limits appear. The use of OLTCs in the distribution

Fig. 4.83 Common evaluation hexagon of Volt/var control strategies

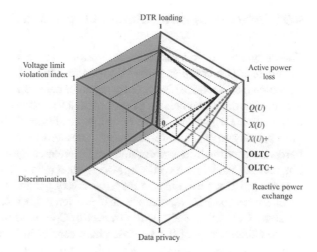

substations results in a considerable voltage limit violation index within the rural LV_Link-Grid. Compared to the other control strategies, it modifies the DTR loading, the active power loss and the MV-LV reactive power exchange only insignificantly. Data privacy is preserved, and the customers are not discriminated. Its combination with Q-Autarkic CPs significantly decreases the MV-LV reactive power exchange and slightly increases the voltage limit violation index. Furthermore, the DTR loading and active power loss are marginally reduced.

Due to its inability to mitigate violations of the lower voltage limit, the $\cos\varphi(P)$-control should not be used for voltage control at the LV level.

Figure 4.83 shows the common evaluation hexagon of Volt/var control strategies. It is calculated by superimposing the results of both LV_Link-Grids (see Sect. A.4.3.2). The $\cos\varphi(P)$-control is excluded from the diagram as it is ineligible for voltage control at the LV level.

The $Q(U)$-control leaves significant voltage limit violations and provokes excessive DTR loading, active power loss and MV-LV reactive power exchange. It does neither support data privacy nor non-discriminatory ancillary service procurement. Compared to $Q(U)$, the OLTC and its combination with Q-Autarkic CPs have better voltage regulation ability. It provokes lower DTR loading, active power loss and MV-LV reactive power exchanges; And guarantees data privacy and freedom from discrimination. $X(U)$-control is the most reliable solution to maintain voltage limit compliance. It eliminates almost all limit violations while provoking relative low DTR loadings, active power losses and MV-LV reactive power exchanges. Its combination with CP_Q-Autarky slightly impairs limit compliance while reducing the MV-LV reactive power exchange significantly.

4.9   Volt/var Control Chain Setup in *Y*-axis

The design of the Volt/var chain control most suitable for a specific vertical power system axis is a complex topic that requires the close integration of planning and operation (see Sect. 2.8). Implementing automation and communication or installing new control devices increases the electrical infrastructure utilisation and causes additional costs. Depending on the prevalent conditions, this may increase or decrease the total system costs. Careful analysis must ensure an adequate trade-off between hardware-, automation- and communication-related expenditures already in the planning stage. Due to each vertical power system axis's distinct characteristics, design recommendations can only be made on a very general level. The general guideline on how to set the Volt/var chain control is given in Sect. 4.9.1. To facilitate the setting process, Sect. 4.9.2 analyses the impact of Compensating Device (CD) placement on the effectiveness of the Volt/var chain control in a theoretical distribution grid. Section 4.9.3 presents the Volt/var chain control setup procedure for the vertical axis, including MV, LV, CP, and consuming device level.

4.9.1   General Guideline

Vertical power system axes are very complex in nature as they include the complete fractal set of smart grids (see Sect. 2.3.3). Their ramified structures interconnect millions of customer plants and numerous producers and storages owned and operated by different stakeholders. Meanwhile, compliance to the operational limits and the optimal use of the existing infrastructures is of utmost importance for utilities. These conditions give rise to several basic requirements on the Volt/var chain control design:

- **Reliability**: The solution shall guarantee voltage limit compliance during normal system operation.
- **Efficiency**: The solution shall provide energy efficiency by dynamic optimisation.
- **Social compatibility**: The solution shall preserve the interests of all involved stakeholders, including non-discriminatory market access and data privacy.
- **Economy**: The solution shall represent an appropriate trade-off between hardware-, automation- and communication-related expenditures.

The Volt/var chain control is derived from the generalised form described in Sect. 4.6.2. The following should be considered:

- **Control the reactive power flow** through external boundary nodes.
  LINK-Smart Grids consists of numerous Links that are owned and operated by different stakeholders. They are interconnected via external boundary nodes and interfaces (see Sect. 2.6.2.1). Each Grid-Link operator should maintain voltage limit compliance in its Grid-Link or Grid-Link-bundle by controlling the reactive power exchanges with the neighbouring external Links.

- **Minimisation of the number of control units**
  The reduction of capital and operational expenditures of the Volt/var chain control is directly related to the number of control units. Based on the fractal dimension analyses (see Sect. 2.3.4), the Smart Grids' realisation requires the highest global resources in CP and LV levels. The focus is minimising the number of control units in those levels by using an appropriate Volt/var control chain strategy.
- **Selection of the appropriate Volt/var control strategy in the Grid-Link chain**
  The selection of an appropriate Volt/var control strategy for each Grid-Link is possible through a holistic analysis of the control chain. The general description of the Volt/var control chain creates the basis for its analyses and design (see Sect. 4.6.2).

4.9.2 Reactive Power Compensation in the Vertical Axis

Generally, the var contributions of shunt-connected RPDs, producers and storages are controlled for two distinct purposes: Voltage support and load compensation. Both aspects must be considered to design the effective **control ensemble**, which maintains voltages with minimal reactive power flows. Distributed and concentrated var contributions are used to maintain voltage limit compliance throughout the grid (see Sect. 4.6.3.1). They provoke reactive power flows that are compensated by elements connected somewhere within the grid. Both the choice between distributed and concentrated var contributions and the placement of the compensating devices significantly affect the resulting effectiveness of the control ensemble.

From a physical point of view, the total reactive power contribution of all elements within an isolated power system is always balanced. As a consequence, the reactive power used to maintain voltages at the LV level must be compensated by other elements connected somewhere within the system. Distinct operators often manage the different grids, so the controllability of the corresponding Q-exchanges is of great interest. The standardised structure of the *LINK*-Architecture allows applying load compensation, which is traditionally used at the device level or in industrial CPs (see Sect. 4.2.3.1), to any Grid-Link within the vertical power system axis, i.e. HV, MV, LV and CP_Grid-Link. The latter case is described in Sect. 4.6.3.3. The impact of load compensation at the MV and LV level on the effectiveness of the resulting Volt/var control ensemble is analysed in this section.

The setups used to study this impact are shown in Figs. 4.84 and 4.85. They include detailed models of an 18.5 MVA STR with its tap changer fixed in mid-position, a 24 km MV cable feeder, and 32 equal LV_Link-Grids, each connecting 61 Q-Autarkic residential CPs. The LV grid is a real rural grid with four feeders where local $X(U)$-control (see Sect. 4.6.3.2) is applied at the two longest feeders. The $X(U)$ local controls provoke the concentrated var contributions required to maintain the voltage at their connection points between 0.91 and 1.09 p.u. The HV grid is modelled as an ideal voltage source (slack) with nominal voltage. All CPs contribute the same active power P^{CP} and no reactive power.

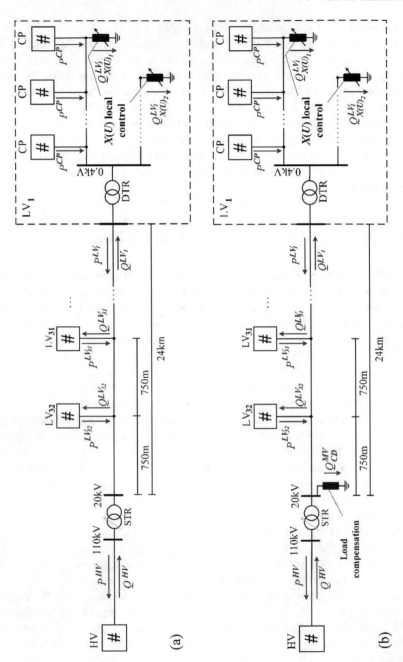

Fig. 4.84 Setups used to analyse the impact of CD placement on the effectiveness of the Volt/var control ensemble: **a** No CDs; **b** CD at STR secondary bus

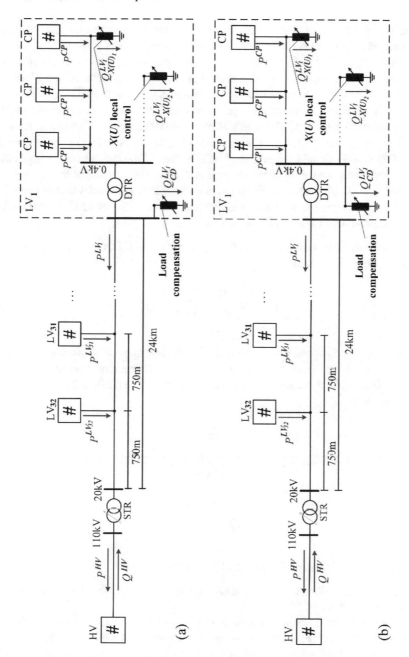

Fig. 4.85 Setups used to analyse the impact of CD placement on the effectiveness of the Volt/var control ensemble: **a** CDs at DTR primary buses; **b** CDs at DTR secondary buses

In Fig. 4.84a, no compensating devices are installed at the LV and MV level that compensate for the reactive power flows. Reactive power is exchanged between HV, MV and LV level. In the other cases, compensating devices are applied that compensate for the reactive power exchanges between the different grid levels. In Fig. 4.84b, the device is connected to the secondary bus of the STR, compensating the reactive power flow at the STR primary side. The compensation of the reactive power exchange between LV and MV level is also investigated. Therefore, in each of the 32 distribution substations, the compensating device is connected to the primary (Fig. 4.85a) and secondary bus (Fig. 4.85b) of the DTR, respectively. In both cases, they compensate for the reactive power flow at the DTR primary side.

Load flow simulations are conducted in each setup for gradually increasing values of P^{CP}. The effectiveness of the resulting Volt/var control ensemble is assessed for the different setups based on three criteria:

- The total reactive power contribution $Q_{X(U)}^{\Sigma}$ of all $X(U)$ local controls, calculated according to Eq. (4.46).

$$Q_{X(U)}^{\Sigma} = \sum_{i=1}^{32} \left(Q_{X(U)_1}^{LV_i} + Q_{X(U)_2}^{LV_i} \right) \qquad (4.46)$$

- The total reactive power contribution Q_{CD}^{Σ} of all compensating devices, calculated according to Eqs. (4.47a), (4.47b) and (4.47c), for the setups without CDs, with one CD at the STR secondary bus, and with one CD in each distribution substation, respectively.

$$Q_{CD}^{\Sigma} = 0 \qquad (4.47a)$$

$$Q_{CD}^{\Sigma} = Q_{CD}^{MV} \qquad (4.47b)$$

$$Q_{CD}^{\Sigma} = \sum_{i=1}^{32} Q_{CD}^{LV_i} \qquad (4.47c)$$

- The voltage profiles of the MV and LV feeders.

Figure 4.86a shows the total reactive power contributions of all compensating devices as functions of P^{CP}.

For active power injections below 2.81 kW per CP, the CD connected at the STR secondary bus consumes reactive power to compensate for the Q-production of the MV cables. For higher P-injections, it produces reactive power to meet the remaining demand of the connected LV grids. The CDs located in the distribution substation produce insignificant amounts of reactive power for $P^{CP} < 2.2$ kW to compensate for the losses at the LV level. Their Q-production drastically increases for greater P-injections, especially if they are connected at the DTR secondary side.

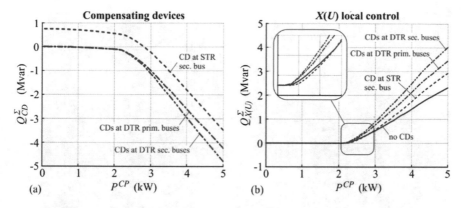

Fig. 4.86 Total reactive power contribution of different control devices: **a** Compensating devices; **b** Voltage maintaining devices

The $X(U)$ local controls consume reactive power to push down the voltages when the active power injection per CP exceeds 2.1 kW, Fig. 4.86b. For P-injections above 2.81 kW, the Q-production of the CDs increases the amount of reactive power required to maintain acceptable voltages, especially if they are placed close to the $X(U)$s. For P-injections between 2.1 and 2.81 kW, the reactive power consumption of the CD located in supplying substation slightly reduces the consumption of the corresponding $X(U)$s.

The voltage profiles of the MV and LV feeders are shown in Figs. 4.87 and 4.88 for the different CD placements and active power injections of 5 kW per CP. A clear trend is observable at first glance: The closer the compensating devices and $X(U)$s are placed together, the lower is the resulting margin to the upper voltage limit. In Figs. 4.87 and 4.88a, the voltage drop over the DTR appears as a discontinuity in the profile, as the DTRs are associate with zero length. When the CDs are located at the DTR secondary buses, the reactive power flows through the DTRs are almost compensated, reducing the corresponding voltage drops significantly. As a result, maintaining acceptable voltages is aggravated. All in all, the following trend is observed:

> The closer the compensating devices and $X(U)$s are placed together, the less effective is the resulting Volt/var control ensemble.

4.9.3   Setup Procedure of the Volt/var Chain Control's Structure

The generalised chain setup procedure is designed to meet the basic requirements presented in Sect. 4.9.1. Due to the major disadvantages of the distributed $\cos\varphi(P)$ and $Q(U)$ controls and the OLTC at the distribution substation identified in Sects. 4.6.3.1, 4.7 and 4.8, only the $X(U)$ control is considered for setting up the Volt/var

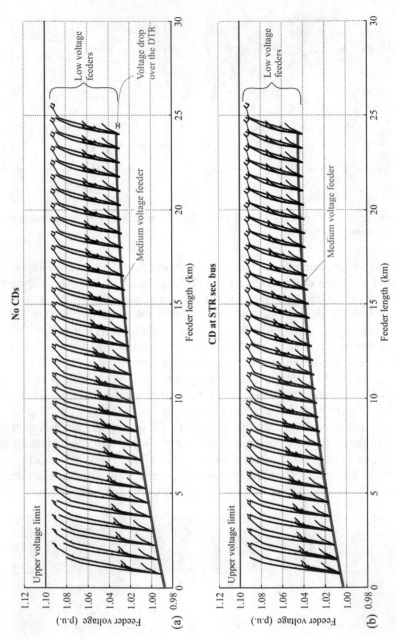

Fig. 4.87 Voltage profiles of MV and LV feeders for $P^{CP} = 5$ kW and different CD placements: **a** No CDs; **b** CD at STR secondary bus

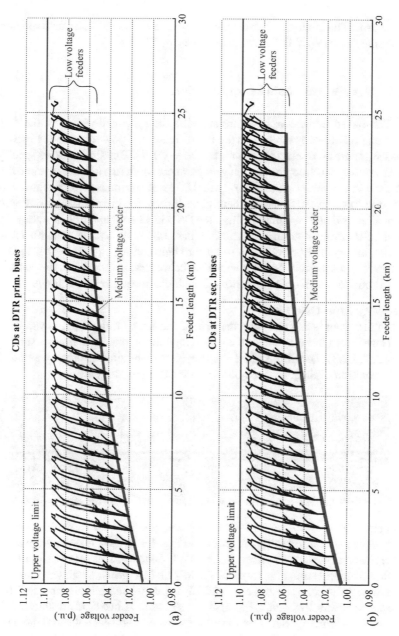

Fig. 4.88 Voltage profiles of MV and LV feeders for $P^{CP} = 5$ kW and different CD placements: **a** CDs at DTR primary buses; **b** CDs at DTR secondary buses

chain control. The chain setup procedure is used to set up the Volt/var chain control in an exemplary MV-LV-CP chain to clarify its application.

4.9.3.1   Generalised Chain Setup Procedure

The *LINK*-based holistic architecture supports the systematic setup of the vertical Volt/var chain control as it relies on the fractal feature of power systems. Figure 4.89 shows the generalised setup procedure for the MV-LV-CP chain. It allows specifying the Volt/var control structure of each grid part according to a bottom-up approach: it starts at the fractal level four of Smart Grids (CP level) and ends at the fractal level two (medium voltage level). Due to the central position of the distribution grids (between transmission grid and customer plants), the DSO has a strategic role in setting up the vertical Volt/var chain control. Therefore, the DSO may use the generalised setup procedure shown in Fig. 4.89 to determine the chain control's structure in the MV-LV-CP chain in accordance with all involved stakeholders, i.e. the TSO, the customers, and the producer and storage operators. It includes seven steps as follows.

1. **Specify Link-Grid Size**

   The first step is to determine the size of the Link-Grid, which is variable and defined from the area where the corresponding secondary control is set up (see Sect. 2.5.2.3). The Link-Grid size is crucial to maintain data privacy, as it defines the position of the interfaces between the different grid parts.

   > The involved stakeholders may wish to define individual objective functions for the *VvSC* of each Link-Grid. For example, environmentalists may set the objective function of their $VvSC^{CP}$ to minimise CO_2 emissions or maximise the PV self-consumption, while money-saving people may prefer to minimise costs or maximise revenues, etc.

   Figure 4.90 shows different Link-Grid arrangements in the vertical axis resulting from the generalised setup procedure. In Fig. 4.90a, the Link-Grids are set upon the classical power system levels, i.e. MV, LV and CP. This setup may be suitable when different companies operate the MV and LV grids. Meanwhile, one Link-Grid may be set upon the MV and LV levels when both are operated by the same company, Fig. 4.90b. Figure 4.90c shows the Link-Grid arrangement in an apartment building. Separate Link-Grids are set upon each flat and the housing area (see Sect. 2.12.2.2), or depending on the circumstances, it can be applied to a housing area.

2. **Specify Control Targets**

   This step dedicates the specification of the Volt/var secondary control targets of the Link-Grid. The potential targets of Volt/var control are:

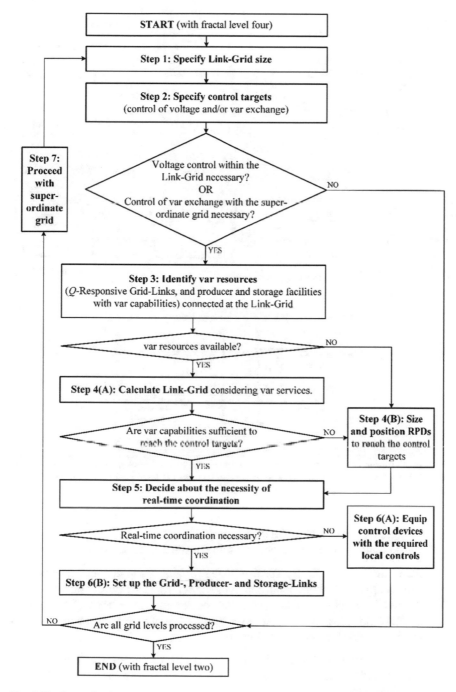

Fig. 4.89 Generalised setup procedure of the Volt/var chain control in the MV-LV-CP chain

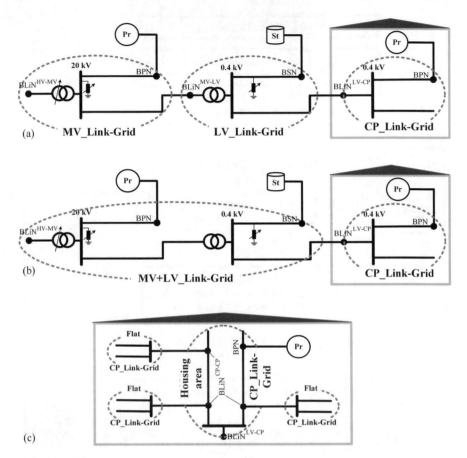

Fig. 4.90 Different Link-Grid arrangements in the vertical axis: **a** Link-Grids are set upon the MV, LV and CP levels; **b** One Link-Grid is set upon the MV and LV levels; **c** Link-Grids are set upon each flat and the housing area of an apartment building

- Voltage control within the own Link-Grid;
- And control of the var exchange with the superordinate grid.

The Grid Codes typically specify voltage limits at the connection points of customer plants. In these conditions, voltage control may be necessary at the HV, MV and LV level but not at the CP and consuming device level. The operators of Link-Grids must agree upon the controllability of their var exchange. *LINK*-Solution offers different control modes:

- *Q*-**Responsive** Grid-Link that respect set-points received from superordinate Grid-Links in real-time. Data exchange between both involved Grid-Links is necessary;
- *Q*-**Autarkic** Grid-Link does not exchange any reactive power with the super-ordinate Link-Grid (see Sect. 4.6.3.3 for the CP level). No data exchange is needed; And

- **Q-Uncontrolled** Link-Grid that provokes uncontrolled reactive power exchanges with the superordinate Link-Grid.

3. **Identify Var Resources**
   This step identifies the available var resources. From the Link-Grid's point of view, var resources include:

   - Producer facilities with var capabilities;
   - Storage facilities with var capabilities;
   - And subordinate Q-Responsive Grid-Links.

   The identified var resources may provide var services to reach the specified control targets.

4. (A) **Calculate Link-Grid**
      If var resources are available, calculate the Link-Grid under the consideration of var services to check whether the available var resources are sufficient to reach the specified control targets. Consider var services procured by non-discriminatory, market-based and transparent procedures to comply with [22]. If no var resources are available, proceed with step 4 (B).

   (B) **Size and position RPDs**
      When no var resources are available, or their var capabilities are insufficient to reach the control targets, RPDs must be sized and positioned within the Link-Grid. The identification of the appropriate size, position and number of RPDs requires careful analysis of the Link-Grid.

5. **Decide About the Necessity of Real-Time Coordination**
   The real-time coordination of the available var resources allows optimising the grid operation but requires distributed data collection, central data processing and communication infrastructure. Analyse the costs and benefits of coordination and decide about its implementation.

6. (A) **Equip control devices with the required local controls**
      If coordination is not necessary, equip the RPDs and the producer and storage facilities with the local Volt/var controls required to reach the control targets.

   (B) **Set up the Grid, Producer- and Storage-Links**
      If coordination is necessary, upgrade the Link-Grid with secondary control and interfaces and each producer and storage facility that possesses var capabilities with primary control and interface.

7. **Proceed with Superordinate Grid**
   If not all grid levels are processed, repeat the generalised setup procedure for the superordinate grid. Otherwise, terminate the procedure.
   An exemplary setup of the Volt/var chain control in the MV-LV-CP levels illustrates the setup procedure in the following.

4.9.3.2 Setup Example

The considered MV-LV-CP chain consists of one MV grid that includes solely one voltage level (e.g. 20 kV), several LV grids, and numerous single-family houses connected at the LV level. One DSO operates the MV and all LV grids. Producer and storage facilities are connected to the MV and CP grids but not to the LV grids. While the STR possesses an OLTC, the DTRs do not.

Application of the setup procedure
Figure 4.91 shows the procedure used to determine the Volt/var chain control's structure for the regarded MV-LV-CP chain. It is directly derived from the generalised one by starting at fractal level four (CP level) and ending at fractal level two (MV level).
Determination of the control structure at the CP level

The first step is to determine the Link-Grid size of all CPs included in the MV-LV-CP chain. In the considered example, all CPs are single-family houses. Therefore, one Link-Grid is set upon each CP; the $BLiN^{LV-CP}$s are the CPs' connection points to the LV grids. Voltage control at the CP level is not required since the distances between the $BLiN^{LV-CP}$s and the connection points of various devices (CP's sockets) are minimal; A significant voltage drop is usually not detectable. However, the customers and the DSO agree to eliminate the uncontrolled LV-CP var exchange by setting the CP_Grid-Link with the parametrisation Q-Autarky. Each CP has one PV and one battery system with var capabilities that are used to achieve the CP_Q-Autarky. The calculation of the CP_Link-Grids shows that the var capabilities of the PV and battery systems are sufficient to compensate the var requirements within the CP fully and, as a result, to eliminate the LV-CP var exchange of each CP. Coordination of the PV and battery systems' var contributions is necessary for each CP to realise Q-Autarky. Both the PV and battery systems are upgraded with primary controls and interfaces, and the CP_Link-Grids are upgraded with secondary controls and interfaces to the corresponding Producer- and Storage-Links. No LV-CP interfaces are implemented as no LV-CP data exchange is necessary for the Q-Autarkic operation mode of CPs.

The resulting Volt/var control structure at the CP level is shown in Fig. 4.92. While the power flow path is shown in black, the automation and communication path is blue-coloured. Golden ellipses indicate Grid-, Producer- and Storage-Links. The $VvSC^{CP}$ calculates real-time var set-points for the primary controls ($PC_{Pr/St}^{CP}$) of the Producer- and Storage-Link to respect the fixed var constraint ($Cns_{NgbLV}^{CP} = 0$ kvar) at the $BLiN^{LV-CP}$. Neither reactive power nor any related data is exchanged between the LV grid and the CPs.

Determination of the control structure at the LV level

Each LV_Link-Grid's size is specified by setting the $BLiN^{MV-LV}$ to the primary busbar of the DTR. Due to the high PV penetration in the CP level, voltage control is necessary at the LV level. Meanwhile, the DSO does not intend to control the MV-LV var exchange for the following reasons:

CP level

Step 1:	Each CP is a single-family house and shall include one Link-Grid. The LV-CP boundary link nodes are set to the CPs' grid connection points
Step 2:	Each CP shall act Q-Autarkic
Step 3:	Each CP includes a PV and battery system with var capabilities
Step 4(A):	The available var capabilities are sufficient to achieve CP_Q-Autarky
Step 5:	The var contributions of the PV and battery system must be coordinated to eliminate the LV-CP var exchange
Step 6(B):	Grid-, Producer- and Storage-Links at the CP level are set up

LV level

Step 1:	The MV-LV boundary link nodes are set to the DTRs' primary bus bars
Step 2:	Voltage control is necessary at the LV level
Step 3:	No var resources are available at the LV level
Step 4(B):	$X(U)$ control devices are installed close to the ends of the violating LV feeders
Step 5:	Due to the low number of var resources, no coordination is necessary at the LV level
Step 6(A):	$X(U)$ control devices are equipped with local control

MV level

Step 1:	The HV-MV boundary link node is set to the STR's primary bus bar
Step 2:	Control of the voltage at the MV level and the HV-MV var exchange is necessary
Step 3:	Run-of-river power plants, wind turbines and battery systems with var capabilities are connected to the MV_Link-Grid
Step 4(A):	The available var capabilities are insufficient to control the HV-MV var exchange within the agreed range
Step 4(B):	An additional CD is positioned at the STR's secondary bus bar
Step 5:	Coordination is necessary to optimally utilise the available var resources
Step 6(B):	Grid-, Producer- and Storage-Links at the MV level are set up

Fig. 4.91 Procedure used to determine the Volt/var chain control's structure for an exemplary MV-LV-CP chain

- The DSO operates both the MV and LV grids. Therefore, all BLiN^{MV-LV} are internal boundary nodes.
- The analysis conducted in Sect. 4.9.2 indicates that compensating for the reactive power flow through the DTR increases the reactive power consumption of the $X(U)$ local controls needed to maintain voltage limit compliance at the LV level.

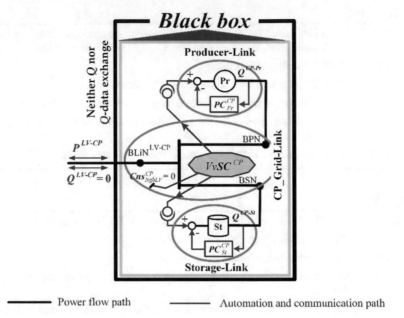

Fig. 4.92 Volt/var control structure at the CP level of the exemplary MV-LV-CP chain (Q-Autarkic CP_Grid-Link)

No var resources are available from the viewpoint of the LV_Link-Grids as no producer and storage facilities are directly connected, and all CPs act Q-Autarkic. Therefore, $X(U)$ control devices are installed close to the ends of the limit violating LV feeders, and their rating is determined based on load flow calculations and worst-case assumptions. Due to the low number of var resources in each LV_Link-Grid and the reliability and effectiveness of $X(U)$ local control, coordination at the LV level is not necessary. Instead, the $X(U)$ control devices are equipped with local control.

The resulting Volt/var control structure at the LV-CP chain is shown in Fig. 4.93. Gold-coloured dashed ellipses indicate Link-Grids. $X(U)$ local controls maintain voltage limit compliance at the LV level by provoking minimal uncontrolled reactive power exchanges between the MV and the LV_Link-Grids. No communication infrastructure is necessary at the LV level as no data is exchanged between the MV and LV and the LV and CP level.

Determination of the control structure at the MV level

The MV_Link-Grid size is specified by setting the BLiN^{HV-MV} to the STR primary busbar and the above determined BLiN^{MV-LV}s. Voltage control within the MV_Link-Grid is necessary to maintain voltage limits compliance. Furthermore, the DSO may agree with TSO to provide reactive power as an ancillary service by controlling the HV-MV var exchange within a specific range. Run-of-river power plants, wind turbines and battery systems with var capabilities are connected to the MV_Link-Grid. However, the calculation of the MV_Link-Grid shows that procuring var

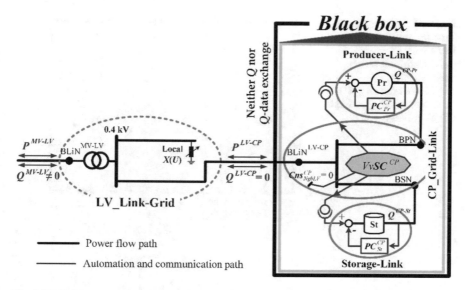

Fig. 4.93 Volt/var control structure at the exemplary LV-CP chain (LV_Link-Grid with $X(U)$ local control and Q-Autarkic CP_Grid-Link)

services from these resources is insufficient to control the agreed HV-MV reactive power exchange; additional CD within the MV_Link-Grid are necessary. In analogy with the conclusion drawn in Sect. 4.9.2, the analysis of the MV_Link-Grid showed that positioning the CD at the secondary busbar of the STR maximises the effectiveness of the resulting Volt/var control chain; the appropriate CD rating is calculated. In agreement with the relevant stakeholders (for example, producer operators, etc.), the DSO implements MV_Link-Grid at the MV level to optimise the operational grid performance and enable non-discriminatory, market-based, and transparent procurement var services. Therefore, the necessary automation and communication are implemented as follows:

- The OLTC of the STR, which already possesses primary control;
- The CD is upgraded with primary control;
- The producer and storage facilities are upgraded with primary controls and interfaces; and.
- The MV_Link-Grid is upgraded with Volt/var secondary control and interfaces to the Producer- and Storage-Links and to the HV grid.

The resulting Volt/var control structure at the MV-LV-CP chain is shown in Fig. 4.94. $VvSC^{MV}$ coordinates the var contributions of the CD and the Producer- and Storage-Links with the OLTC of the STR by:

- Calculating real-time var set-points for the CDs and the Producer- and Storage-Links connected to the MV_Grid-Link;
- Calculating real-time voltage set-points for the OLTC of the STR;
- While respecting the internal voltage limits and the dynamic var constraint at the $BLiN^{HV-MV}$.

Fig. 4.94 Volt/var control structure at the exemplary MV-LV-CP chain (MV_Grid-Link with additional CD, LV_Link-Grid with $X(U)$ local control and Q-Autarkic CP_Grid-Link)

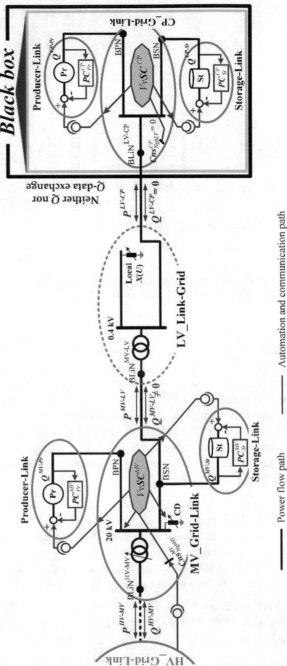

4.10 Setup overview

The Volt/var chain control setup shown in Fig. 4.94 is designed to meet the basic requirements by following the general guideline (see Sect. 4.9.1). The necessity of the $VvSC^{LV}$ is eliminated by combining the $X(U)$ local control with the Q-Autarky of customer plants. This control ensemble maintains voltage limit compliance at the LV level effectively (see Sect. 4.6.3.1) without requiring any data exchanges between the DSO and the customers. The uncontrolled reactive power flows are reduced to their practical minimum. The $VvSC^{MV}$ coordinates the var resources at the MV level with the OLTC of the STR to control the voltage at the MV level and respect var constraints at the BLiN^{HV-MV}. Equation (4.48) compactly represents the control variables and dynamic constraints of the designed vertical Volt/var chain control.

$$
\begin{aligned}
VvSC_{Chain}^{MV-LV-CP} = \Big\{ & VvSC^{MV}\left(PC_{OLTC}^{MV}, PC_{Pr}^{MV}, PC_{St}^{MV}, PC_{CD}^{MV}; Cns_{NgbHV}^{MV}\right), \\
& LC_{X(U)}^{LV}, \\
& VvSC^{CP}\left(PC_{Pr}^{CP}, PC_{St}^{CP}; Cns_{NgbLV}^{CP} = 0\text{kvar}\right) \Big\}
\end{aligned}
\tag{4.48}
$$

4.10.1 MV level

The $VvSC^{MV}$ calculates in real-time:

- The voltage set-points for the primary controls PC_{OLTC}^{MV} of the supplying transformers and other transformers included in the MV_Grid-Link that have OLTC;
- The voltage and reactive power set-points for the primary controls PC_{Pr}^{MV} of the Producer-Links connected to the MV_Grid-Link;
- The voltage and reactive power set-points for the primary controls PC_{St}^{MV} of the Storage-Links connected to the MV_Grid-Link;
- And the voltage, reactive power and switch position set-points for the primary PC_{RPD}^{MV} and direct controls DiC_{RPD}^{MV} of the RPDs included in the MV_Grid-Link;

While respecting:

- The reactive power constraints Cns_{NgbHV}^{MV} at the boundary node to the neighbouring HV_Grid-Link.

4.10.2 LV level

No $VvSC^{LV}$ is used. The voltage at the LV level is controlled by $X(U)$ local control (see Sect. 4.6.3.2).

4.10.3   CP level

The $VvSC^{CP}$ realises CP_Q-Autarky (see Sect. 4.6.3.3) by calculating in real-time:

- The voltage and reactive power set-points for the primary controls PC_{Pr}^{CP} of the Producer-Links connected to the CP_Grid-Link;
- And the voltage and reactive power set-points for the primary controls PC_{St}^{CP} of the Storage-Links connected to the CP_Grid-Link;

Appendix

A.4.1 Behaviour of Different Link-Grids

This appendix catalogues the simulated behaviour of all Link-Grids listed in Table 4.5 under different Volt/var control strategies and for the case without any Volt/var control. The rural residential CP_Link-Grid and the rural LV_Link-Grid are calculated for both spiky and smooth load profiles at the CP level. Meanwhile, all other Link-Grids are simulated only for the smoothed load profiles.

Table 4.5   Overview of the simulated Link-Grids

Level	Link-Grid	Load profiles at the CP level
CP	Rural residential	Spiky
		Smooth
	Urban residential	Smooth
LV	Rural	Spiky
		Smooth
	Urban	Smooth
MV	Small	Smooth
	Large	Smooth

A.4.1.1 Spiky Load Profiles

CP level

Rural residential CP_Link-Grid connected to LV level

The model of the rural residential CP_Link-Grid is specified in Fig. 4.95; Its behaviour without any Volt/var control is depicted in Fig. 4.96; And its behaviour with the different Volt/var controls is shown in Fig. 4.97. Furthermore, Fig. 4.98 presents the behaviour of the rural residential CP_Link-Grid for cases A, B, and C.

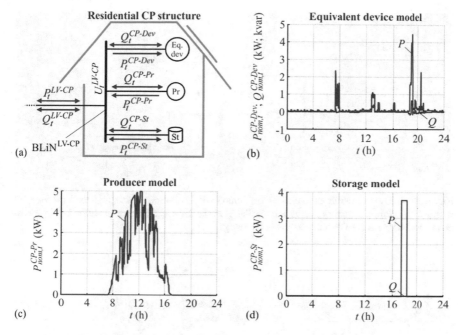

Fig. 4.95 Rural residential CP_Link-Grid: **a** Structure; **b** Spiky load profiles of the Eq. dev.-model; **c** Spiky load profile of the Pr.-model; **d** Spiky load profiles of the St.-model

CP structure	Equivalent device model	SMPS, motors, resistive, lighting
	Producer model	One photovoltaic system
	Storage model	One electric vehicle charger
Equivalent device model	Daily energy consumption	4.25 kWh
	Max. active power consumption	4.46 kW
	Max. reactive power consumption	0.65 kvar
	Max. reactive power production	0.16 kvar
	ZIP coefficients	Time-invariant
	References	[10, 37]
Producer model	Daily energy production	23.14 kWh
	Max. active power production	5.00 kW
	Reactive power contribution	According to Volt/var control strategy
	References	[41]
Storage model	Daily energy consumption	1.85 kWh
	Max. active power consumption	3.7 kW
	ZIP coefficients	Time-invariant
	References	[37, 64]

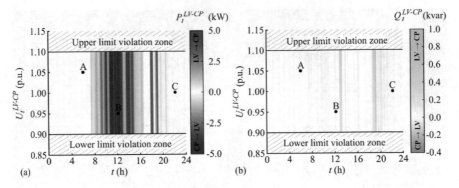

Fig. 4.96 Daily behaviour of the rural residential CP_Link-Grid with spiky load profiles and without any Volt/var control for various voltages at the LV-CP boundary node: **a** LV-CP active power exchange; **b** LV-CP reactive power exchange

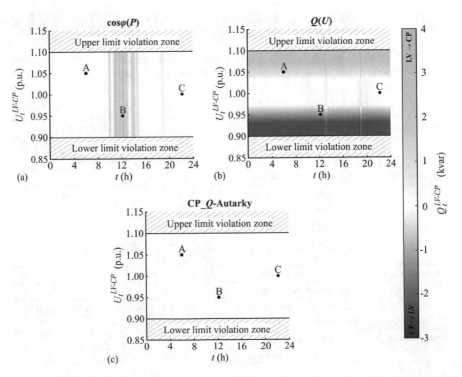

Fig. 4.97 Daily LV-CP reactive power exchange of the rural residential CP_Link-Grid with spiky load profiles for various voltages at the LV-CP boundary node and different control strategies: **a** $\cos\varphi(P)$; **b** $Q(U)$; **c** CP_Q-Autarky

Control strategy	Q_t^{LV-CP} (kvar)		
	Case A	Case B	Case C
None	−0.0044	0.0182	−0.0131
$\cos\varphi(P)$	−0.0044	2.4395	−0.0131
$Q(U)$	1.2056	−1.1918	−0.0131
CP_Q-Autarky	0.0000	0.0000	0.0000

Control strategy	Q_t^{CP-Dev} (kvar)		
	Case A	Case B	Case C
None	−0.0044	0.0182	−0.0131
$\cos\varphi(P)$	−0.0044	0.0182	−0.0131
$Q(U)$	−0.0044	0.0182	−0.0131
CP_Q-Autarky	−0.0044	0.0182	−0.0131

Control strategy	Q_t^{CP-Pr} (kvar)		
	Case A	Case B	Case C
None	0.0000	0.0000	0.0000
$\cos\varphi(P)$	0.0000	2.4213	0.0000
$Q(U)$	1.2100	−1.2100	0.0000
CP_Q-Autarky	0.0044	−0.0182	0.0131

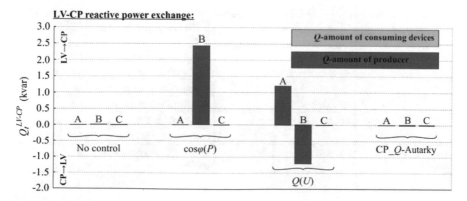

Fig. 4.98 Composition of the LV-CP reactive power exchange of the rural residential CP_Link-Grid with spiky load profiles for different cases, no control and various control strategies

LV level

Rural LV_Link-Grid

The model of the rural LV_Link-Grid is specified in Fig. 4.99; Its behaviour without any Volt/var control is depicted in Figs. 4.100 and 4.101; And its behaviour with the different Volt/var controls is shown in Figs. 4.102 to 4.104. Furthermore, Figs. 4.105 to 4.107 present the behaviour of the rural LV_Link-Grid for cases A, B, and C.

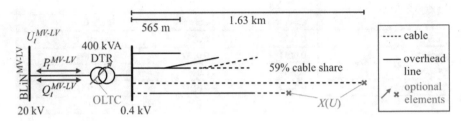

Fig. 4.99 Simplified one-line diagram of the rural LV_Link-Grid (real Austrian grid)

DTR	Rating:		400 kVA	
	Nominal voltage	Primary	21.0 kV	
		Secondary	0.42 kV	
	Short circuit voltage	Total	3.7%	
		Resistive part	1.0%	
Feeders	Nominal voltage		0.4 kV	
	Number of feeders		4	
	Total line length		6.335 km	
	Total cable share		58.64%	
	Feeder length	Maximal	1.630 km	
		Minimal	0.565 km	
Control parameters	OLTC		Upper voltage limit	0.990 p.u
			Lower voltage limit	0.950 p.u
			Min./mid/max. tap positions	1/3/5
			Additional voltage per tap	2.5%
	$X(U)$		Upper voltage limit	1.09 p.u
			Lower voltage limit	0.91 p.u
Connected lumped models	Link-Grids	CP	Residential with spiky load profiles	61

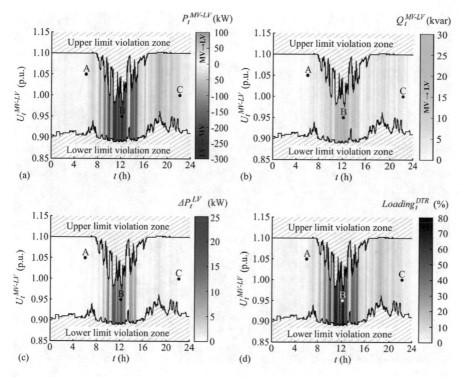

Fig. 4.100 Daily behaviour of the rural LV_Link-Grid without any Volt/var control for various voltages at the MV-LV boundary node and spiky load profiles at the CP level: **a** MV-LV active power exchange; **b** MV-LV reactive power exchange; **c** LV active power loss; **d** DTR loading

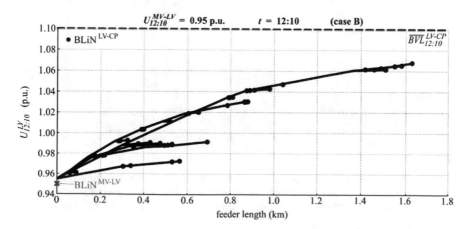

Fig. 4.101 Voltage profiles of the rural LV_Link-Grid's feeders without any Volt/var control at 12:10 for an MV-LV boundary voltage of 0.95 p.u. (case B) and spiky load profiles at the CP level

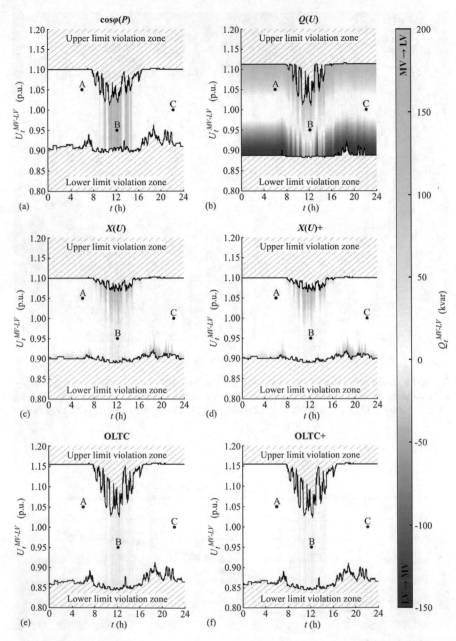

Fig. 4.102 Daily MV-LV reactive power exchange of the rural LV_Link-Grid for various voltages at the MV-LV boundary node, spiky load profiles at the CP level, and different control strategies: **a** $\cos\varphi(P)$; **b** $Q(U)$; **c** $X(U)$; **d** $X(U)$ and CP_Q-Autarky; **e** OLTC; **f** OLTC and CP_Q-Autarky

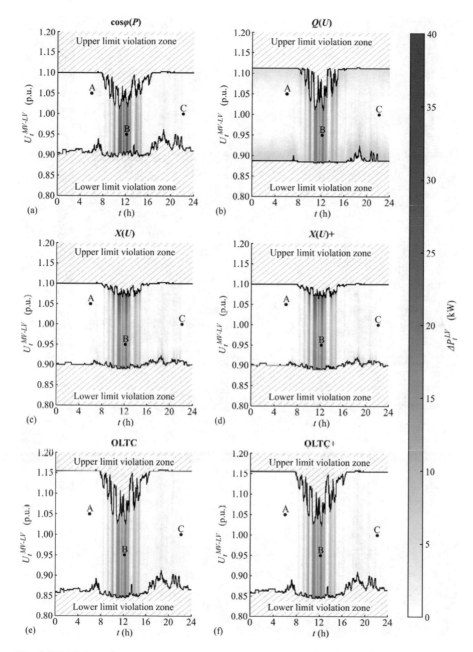

Fig. 4.103 Daily active power loss within the rural LV_Link-Grid for various voltages at the MV-LV boundary node, spiky load profiles at the CP level, and different control strategies: **a** cosφ(P); **b** Q(U); **c** X(U); **d** X(U) and CP_Q-Autarky; **e** OLTC; **f** OLTC and CP_Q-Autarky

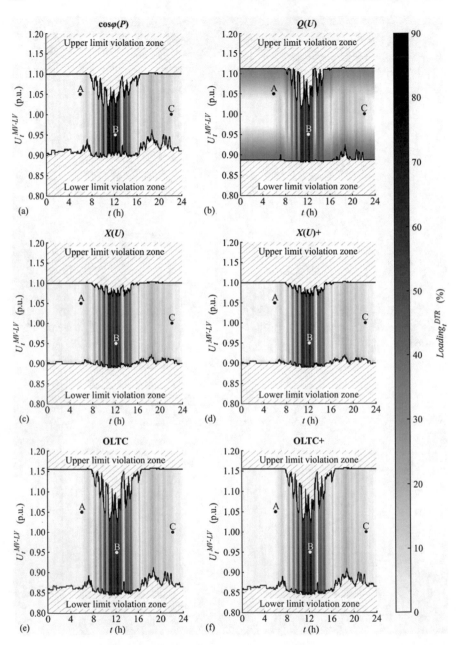

Fig. 4.104 Daily DTR loading within the rural LV_Link-Grid for various voltages at the MV-LV boundary node, spiky load profiles at the CP level, and different control strategies: **a** $\cos\varphi(P)$; **b** $Q(U)$; **c** $X(U)$; **d** $X(U)$ and CP_Q-Autarky; **e** OLTC; **f** OLTC and CP_Q-Autarky

Control strategy	Q_t^{MV-LV} (kvar)		
	Case A	Case B	Case C
None	1.1849	23.7847	3.7581
$\cos\varphi(P)$	1.1849	178.2922	3.7581
$Q(U)$	37.8057	36.1168	3.7581
$X(U)$	1.1849	23.7847	3.7583
$X(U)+$	−0.1576	16.6295	0.0536
OLTC	1.0427	23.4057	3.5752
OLTC+	−0.1403	16.6295	0.0682

Control strategy	$Q_{\Sigma,t}^{LV-CP}$ (kvar)		
	Case A	Case B	Case C
None	1.3421	7.0678	3.6998
$\cos\varphi(P)$	1.3421	154.0554	3.6998
$Q(U)$	37.6942	19.1260	3.6998
$X(U)$	1.3421	7.0678	3.6998
$X(U)+$	0.0000	0.0000	0.0000
OLTC	1.1826	7.4946	3.5024
OLTC+	0.0000	0.0000	0.0000

Control strategy	$Q_{\Sigma,t}^{LV}$ (kvar)		
	Case A	Case B	Case C
None	−0.1572	16.7169	0.0585
$\cos\varphi(P)$	−0.1572	24.2372	0.0585
$Q(U)$	0.1115	16.9907	0.0585
$X(U)$	−0.1572	16.7169	0.0585
$X(U)+$	−0.1576	16.6295	0.0536
OLTC	−0.1399	15.9111	0.0729
OLTC+	−0.1403	16.6295	0.0682

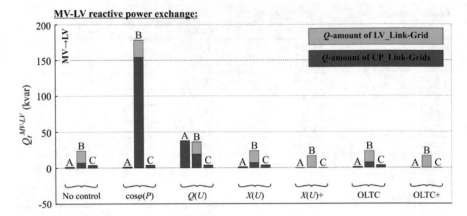

Fig. 4.105 Composition of the MV-LV reactive power exchange of the rural LV_Link-Grid for spiky load profiles at the CP level, different cases, no control and various control strategies

Control strategy	ΔP_t^{LV} (kW)		
	Case A	Case B	Case C
None	0.0250	17.1154	0.2402
$\cos\varphi(P)$	0.0250	24.9791	0.2402
$Q(U)$	0.2931	17.6844	0.2402
$X(U)$	0.0250	17.1154	0.2402
$X(U)+$	0.0245	17.0116	0.2350
OLTC	0.0254	16.3134	0.2475
OLTC+	0.0250	17.0116	0.2426

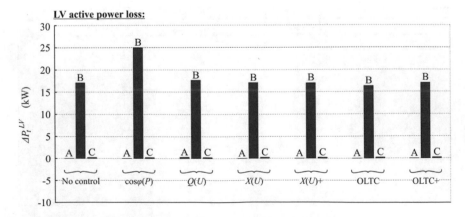

Fig. 4.106 Active power loss within the rural LV_Link-Grid for spiky load profiles at the CP level, different cases, no control and various control strategies

Control strategy	$Loading_t^{DTR}(\%)$		
	Case A	Case B	Case C
None	1.9979	68.0621	7.2600
$\cos\varphi(P)$	1.9979	80.9169	7.2600
$Q(U)$	9.2271	68.3151	7.2600
$X(U)$	1.9979	68.0621	7.2600
$X(U)+$	1.9786	67.9281	7.2011
OLTC	1.9188	68.1371	7.1809
OLTC+	1.9033	67.9281	7.1269

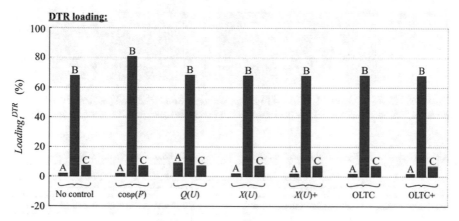

Fig. 4.107 DTR loading within the rural LV_Link-Grid for spiky load profiles at the CP level, different cases, no control and various control strategies

A.4.1.2 Smoothed Load Profiles

CP level

Rural residential CP_Link-Grid connected to LV level

The model of the rural residential CP_Link-Grid is specified in Fig. 4.108; Its behaviour without any Volt/var control is depicted in Fig. 4.109; And its behaviour with the different Volt/var controls is shown in Fig. 4.110. Furthermore, Fig. 4.111 present the behaviour of the rural residential CP_Link-Grid for cases A, B, and C.

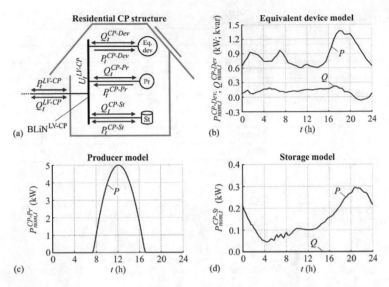

Fig. 4.108 Rural residential CP_Link-Grid: **a** Structure; **b** Smoothed load profiles of the Eq. dev.-model; **c** Smoothed load profile of the Pr.-model; **d** Smoothed load profiles of the St.-model

CP structure	Equivalent device model	SMPS, motors, resistive, lighting
	Producer model	One photovoltaic system
	Storage model	One electric vehicle charger
Equivalent device model	Daily energy consumption	20.75 kWh
	Max. active power consumption[*]	1.37 kW
	Max. reactive power consumption	0.22 kvar
	Max. reactive power production	0.07 kvar
	ZIP coefficients	Time-variant
	References	[58]
Producer model	Daily energy production	31.37 kWh
	Max. active power production	5.00 kW
	Reactive power contribution	According to Volt/var control strategy
	References	[41]
Storage model	Daily energy consumption	3.51 kWh
	Max. active power consumption	0.30 kW
	ZIP coefficients	Time-invariant
	References	[6, 64]

[a]This value is derived from the maximal active power flow through the DTR of the rural LV_Link-Grid measured throughout 2016

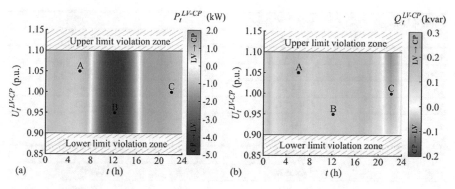

Fig. 4.109 Daily behaviour of the rural residential CP_Link-Grid with smoothed load profiles and without any Volt/var control for various voltages at the LV-CP boundary node: **a** LV-CP active power exchange; **b** LV-CP reactive power exchange

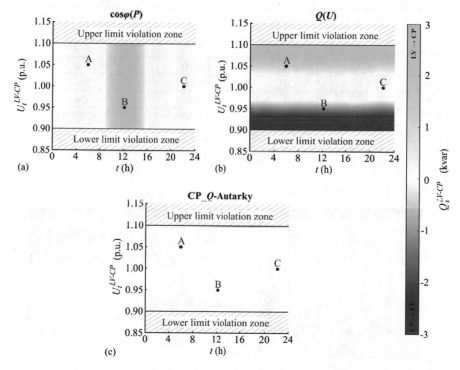

Fig. 4.110 Daily LV-CP reactive power exchange of the rural residential CP_Link-Grid with smoothed load profiles for various voltages at the LV-CP boundary node and different control strategies: **a** $\cos\varphi(P)$; **b** $Q(U)$; **c** CP_Q-Autarky

Control strategy	Q_t^{LV-CP} (kvar)		
	Case A	Case B	Case C
None	0.1791	0.1406	−0.0656
$\cos\varphi(P)$	0.1791	2.5622	−0.0656
$Q(U)$	1.3891	−1.0694	−0.0656
CP_Q-Autarky	0.0000	0.0000	0.0000

Control strategy	Q_t^{CP-Dev} (kvar)		
	Case A	Case B	Case C
None	0.1791	0.1406	−0.0656
$\cos\varphi(P)$	0.1791	0.1406	−0.0656
$Q(U)$	0.1791	0.1406	−0.0656
CP_Q-Autarky	0.1791	0.1406	−0.0656

Control strategy	Q_t^{CP-Pr} (kvar)		
	Case A	Case B	Case C
None	0.0000	0.0000	0.0000
$\cos\varphi(P)$	0.0000	2.4215	0.0000
$Q(U)$	1.2100	−1.2100	0.0000
CP_Q-Autarky	-0.1791	−0.1406	0.0656

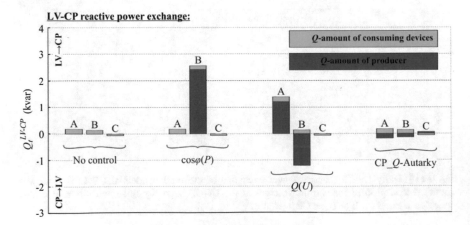

Fig. 4.111 Composition of the LV-CP reactive power exchange of the rural residential CP_Link-Grid with smoothed load profiles for different cases, no control and various control strategies

Urban residential CP_Link-Grid connected to LV level.

The model of the urban residential CP_Link-Grid is specified in Fig. 4.112; Its behaviour without any Volt/var control is depicted in Fig. 4.113; And its behaviour with the different Volt/var controls is shown in Fig. 4.114. Furthermore, Fig. 4.115 present the behaviour of the urban residential CP_Link-Grid for cases A, B, and C.

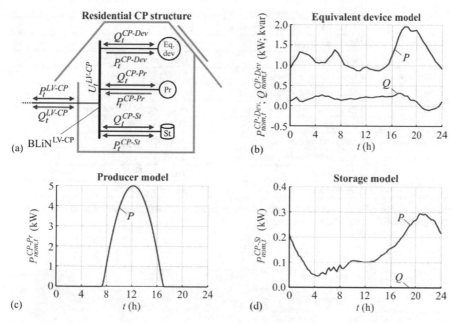

Fig. 4.112 Urban residential CP_Link Grid: **a** Structure; **b** Smoothed load profiles of the Eq. dev.-model; **c** Smoothed load profile of the Pr.-model; **d** Smoothed load profiles of the St.-model

CP structure	Equivalent device model	SMPS, motors, resistive, lighting
	Producer model	One photovoltaic system
	Storage model	One electric vehicle charger
Equivalent device model	Daily energy consumption	29.73 kWh
	Max. active power consumption[a]	1.96 kW
	Max. reactive power consumption	0.32 kvar
	Max. reactive power production	0.10 kvar
	ZIP coefficients	Time-variant
	References	[58]
Producer model	Daily energy production	31.37 kWh
	Max. active power production	5.00 kW
	Reactive power contribution	According to Volt/var control strategy
	References	[41]
Storage model	Daily energy consumption	3.51 kWh
	Max. active power consumption	0.30 kW
	ZIP coefficients	Time-invariant
	References	[6, 64]

[a]This value is derived from the maximal active power flow through the DTR of the urban LV_Link-Grid measured throughout 2016

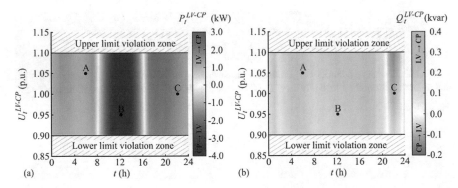

Fig. 4.113 Daily behaviour of the urban residential CP_Link-Grid with smoothed load profiles and without any Volt/var control for various voltages at the LV-CP boundary node: **a** LV-CP active power exchange; **b** LV-CP reactive power exchange

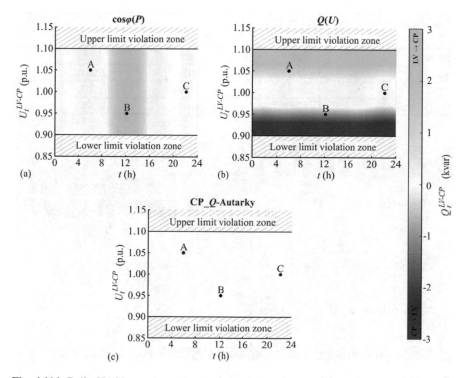

Fig. 4.114 Daily LV-CP reactive power exchange of the urban residential CP_Link-Grid with smoothed load profiles for various voltages at the LV-CP boundary node and different control strategies: **a** $\cos\varphi(P)$; **b** $Q(U)$; **c** CP_Q-Autarky

Control strategy	Q_t^{LV-CP} (kvar)		
	Case A	Case B	Case C
None	0.2566	0.2015	−0.0940
$\cos\varphi(P)$	0.2566	2.6230	−0.0940
$Q(U)$	1.4666	−1.0085	−0.0940
CP_Q-Autarky	0.0000	0.0000	0.0000

Control strategy	Q_t^{CP-Dev} (kvar)		
	Case A	Case B	Case C
None	0.2566	0.2015	−0.0940
$\cos\varphi(P)$	0.2566	0.2015	−0.0940
$Q(U)$	0.2566	0.2015	−0.0940
CP_Q-Autarky	0.2566	0.2015	−0.0940

Control strategy	Q_t^{CP-Pr} (kvar)		
	Case A	Case B	Case C
None	0.0000	0.0000	0.0000
$\cos\varphi(P)$	0.0000	2.4215	0.0000
$Q(U)$	1.2100	−1.2100	0.0000
CP_Q-Autarky	−0.2566	−0.2015	0.0940

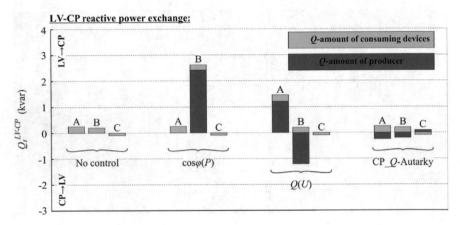

Fig. 4.115 Composition of the LV-CP reactive power exchange of the urban residential CP_Link-Grid with smoothed load profiles for different cases, no control and various control strategies

Commercial CP_Link-Grid connected to MV level.

The model of the commercial CP_Link-Grid is specified in Fig. 4.116; Its behaviour without any Volt/var control is depicted in Fig. 4.117; And its behaviour with the different Volt/var controls is shown in Fig. 4.118. Furthermore, Fig. 4.119 present the behaviour of the commercial CP_Link-Grid for cases A, B, and C.

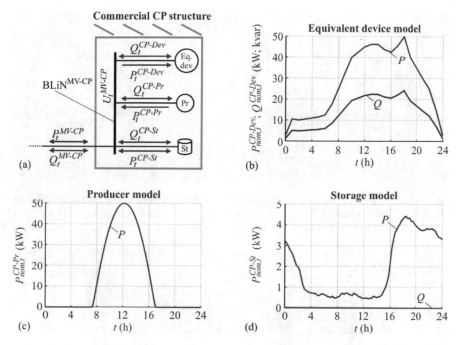

Fig. 4.116 Commercial CP_Link-Grid: **a** Structure; **b** Smoothed load profiles of the Eq. dev.-model; **c** Smoothed load profile of the Pr.-model; **d** Smoothed load profiles of the St.-model

CP structure	Equivalent device model	SMPS, motors, resistive, lighting
	Producer model	One photovoltaic system
	Storage model	Three electric vehicle chargers
Equivalent device model	Daily energy consumption	690.25 kWh
	Max. active power consumption	50 kW
	Power factor	0.90 inductive
	ZIP coefficients	Time-invariant
	References	[12] and [10]
Producer model	Daily energy production	313.65 kWh
	Max. active power production	50 kW
	Reactive power contribution	According to Volt/var control strategy
	References	[41]
Storage model	Daily energy consumption	46.03 kWh
	Max. active power consumption	4.41 kW
	ZIP coefficients	Time-invariant
	References	[6, 64]

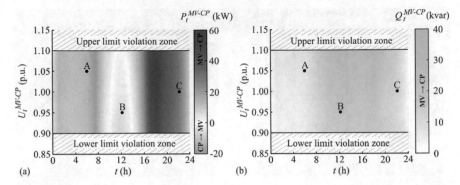

Fig. 4.117 Daily behaviour of the commercial CP_Link-Grid with smoothed load profiles and without any Volt/var control for various voltages at the MV-CP boundary node: **a** MV-CP active power exchange; **b** MV-CP reactive power exchange

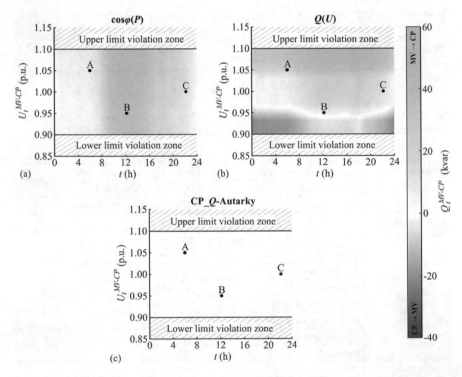

Fig. 4.118 Daily MV-CP reactive power exchange of the commercial CP_Link-Grid with smoothed load profiles for various voltages at the MV-CP boundary node and different control strategies: **a** $\cos\varphi(P)$; **b** $Q(U)$; **c** CP_Q-Autarky

Control strategy	Q_t^{MV-CP} (kvar)		
	Case A	Case B	Case C
None	6.9423	19.5685	11.3008
$\cos\varphi(P)$	6.9423	43.7842	11.3008
$Q(U)$	19.0423	7.4685	11.3008
CP_Q-Autarky	0.0000	0.0000	0.0000

Control strategy	Q_t^{CP-Dev} (kvar)		
	Case A	Case B	Case C
None	6.9423	19.5685	11.3008
$\cos\varphi(P)$	6.9423	19.5685	11.3008
$Q(U)$	6.9423	19.5685	11.3008
CP_Q-Autarky	6.9423	19.5685	11.3008

Control strategy	Q_t^{CP-Pr} (kvar)		
	Case A	Case B	Case C
None	0.0000	0.0000	0.0000
$\cos\varphi(P)$	0.0000	24.2158	0.0000
$Q(U)$	12.1000	−12.1000	0.0000
CP_Q-Autarky	−6.9423	−19.5685	−11.3008

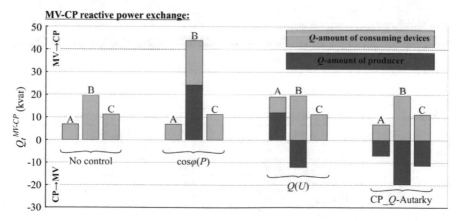

Fig. 4.119 Composition of the MV-CP reactive power exchange of the commercial CP_Link-Grid with smoothed load profiles for different cases, no control and various control strategies

Industrial CP_Link-Grid connected to MV level

The model of the industrial CP_Link-Grid is specified in Fig. 4.120; Its behaviour without any Volt/var control is depicted in Fig. 4.121; And its behaviour with the different Volt/var controls is shown in Fig. 4.122. Furthermore, Fig. 4.123 present the behaviour of the industrial CP_Link-Grid for cases A, B, and C.

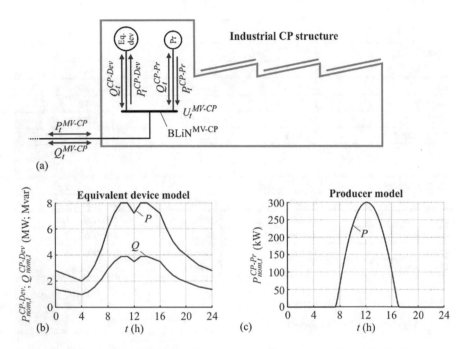

Fig. 4.120 Industrial CP_Link-Grid: **a** Structure; **b** Smoothed load profiles of the Eq. dev.-model; **c** Smoothed load profile of the Pr.-model

CP structure	Equivalent device model	SMPS, motors, resistive, lighting
	Producer model	One photovoltaic system
Equivalent device model	Daily energy consumption	118.44 MWh
	Max. active power consumption[a]	8 MW
	Power factor	0.90 inductive
	ZIP coefficients	Time-invariant
	References	[12] and [10]
Producer model	Daily energy production	1.88 MWh
	Max. active power production	300 kW
	Reactive power contribution	According to Volt/var control strategy
	References	[41]

[a]This value corresponds to the billing demand of the two industrial customers connected to the large MV_Link-Grid

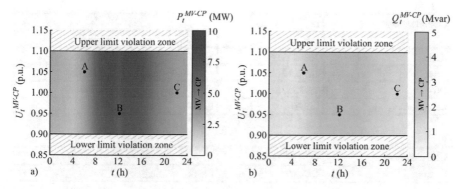

Fig. 4.121 Daily behaviour of the industrial CP_Link-Grid with smoothed load profiles and without any Volt/var control for various voltages at the MV-CP boundary node: **a** MV-CP active power exchange; **b** MV-CP reactive power exchange

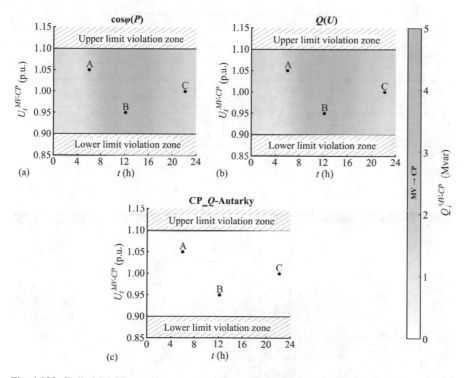

Fig. 4.122 Daily MV-CP reactive power exchange of the industrial CP_Link-Grid with smoothed load profiles for various voltages at the MV-CP boundary node and different control strategies: **a** $\cos\varphi(P)$; **b** $Q(U)$; **c** CP_Q-Autarky

Control strategy	Q_t^{MV-CP} (kvar)		
	Case A	Case B	Case C
None	1676.7231	3267.1158	1518.3498
$\cos\varphi(P)$	1676.7231	3412.4081	1518.3498
$Q(U)$	1749.3231	3194.5158	1518.3498
CP_Q-Autarky	0.0000	0.0000	0.0000

Control strategy	Q_t^{CP-Dev} (kvar)		
	Case A	Case B	Case C
None	1676.7231	3267.1158	1518.3498
$\cos\varphi(P)$	1676.7231	3267.1158	1518.3498
$Q(U)$	1676.7231	3267.1158	1518.3498
CP_Q-Autarky	1676.7231	3267.1158	1518.3498

Control strategy	Q_t^{CP-Pr} (kvar)		
	Case A	Case B	Case C
None	0.0000	0.0000	0.0000
$\cos\varphi(P)$	0.0000	145.2922	0.0000
$Q(U)$	72.6000	−72.6000	0.0000
CP_Q-Autarky	−1676.7231	−3267.1158	−1518.3498

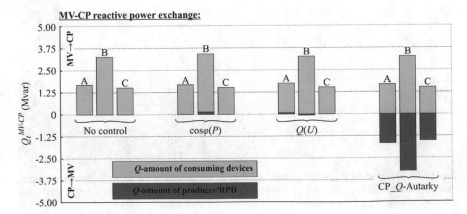

Fig. 4.123 Composition of the MV-CP reactive power exchange of the industrial CP_Link-Grid with smoothed load profiles for different cases, no control and various control strategies

LV level

Rural LV_Link-Grid

The model of the rural LV_Link-Grid is specified in Fig. 4.124; Its behaviour without any Volt/var control is depicted in Figs. 4.125 and 4.126; And its behaviour with the different Volt/var controls is shown in Figs. 4.127 to 4.129. Furthermore, Figs. 4.130 to 4.132 present the behaviour of the rural LV_Link-Grid for cases A, B, and C.

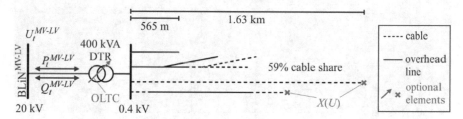

Fig. 4.124 Simplified one-line diagram of the rural LV_Link-Grid (real Austrian grid)

DTR	Rating:		400 kVA
	Nominal voltage	Primary	21.0 kV
		Secondary	0.42 kV
	Short circuit voltage	Total	3.7%
		Resistive part	1.0%
Feeders	Nominal voltage		0.4 kV
	Number of feeders		4
	Total line length		6.335 km
	Total cable share		58.64%
	Feeder length	Maximal	1.630 km
		Minimal	0.565 km
Control parameters	OLTC	Upper voltage limit	0.990 p.u
		Lower voltage limit	0.950 p.u
		Min./mid/max. tap positions	1/3/5
		Additional voltage per tap	2.5%
	$X(U)$	Upper voltage limit	1.09 p.u
		Lower voltage limit	0.91 p.u
Connected lumped models	Link-Grids	CP	Rural residential with smoothed load profiles	61

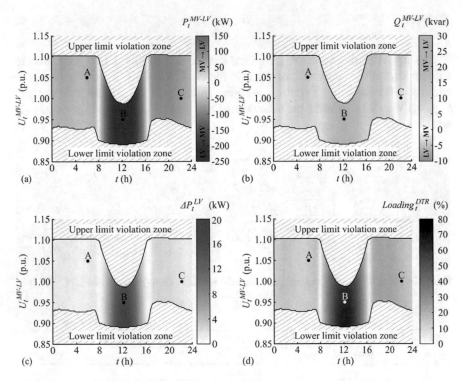

Fig. 4.125 Daily behaviour of the rural LV_Link-Grid without any Volt/var control for various voltages at the MV-LV boundary node and smoothed load profiles at the CP level: **a** MV-LV active power exchange; **b** MV-LV reactive power exchange; **c** LV active power loss; **d** DTR loading

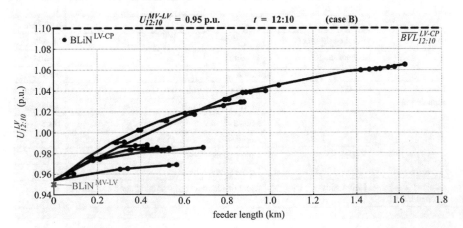

Fig. 4.126 Voltage profiles of the rural LV_Link-Grid's feeders without any Volt/var control at 12:10 for an MV-LV boundary voltage of 0.95 p.u. (case B) and smoothed load profiles at the CP level

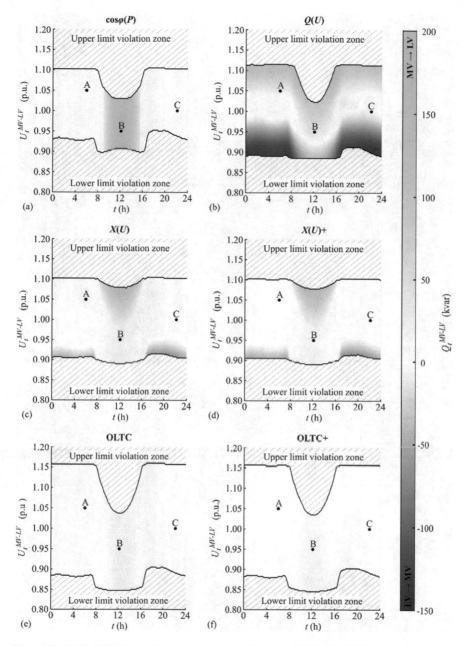

Fig. 4.127 Daily MV-LV reactive power exchange of the rural LV_Link-Grid for various voltages at the MV-LV boundary node, smoothed load profiles at the CP level, and different control strategies: **a** $\cos\varphi(P)$; **b** $Q(U)$; **c** $X(U)$; **d** $X(U)$ and CP_Q-Autarky; **e** OLTC; **f** OLTC and CP_Q-Autarky

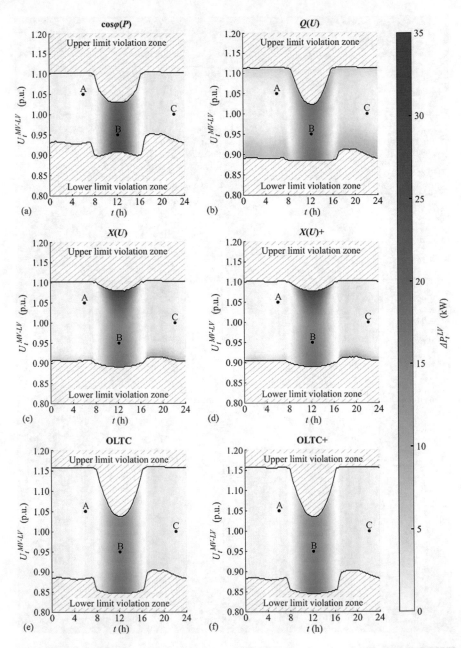

Fig. 4.128 Daily active power loss within the rural LV_Link-Grid for various voltages at the MV-LV boundary node, smoothed load profiles at the CP level, and different control strategies: **a** $\cos\varphi(P)$; **b** $Q(U)$; **c** $X(U)$; **d** $X(U)$ and CP_Q-Autarky; **e** OLTC; **f** OLTC and CP_Q-Autarky

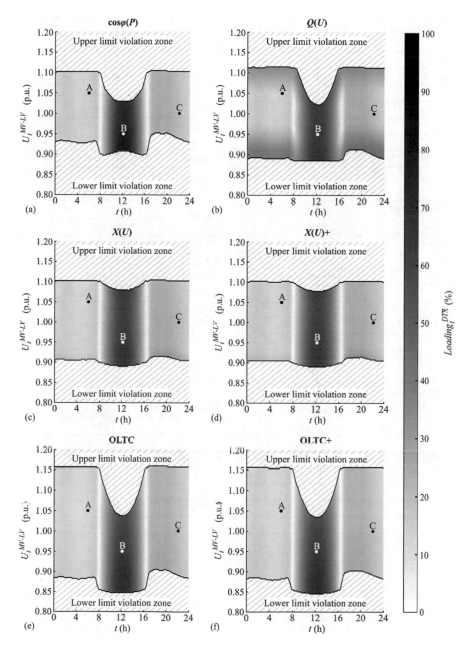

Fig. 4.129 Daily DTR loading within the rural LV_Link-Grid for various voltages at the MV-LV boundary node, smoothed load profiles at the CP level, and different control strategies: **a** $\cos\varphi(P)$; **b** $Q(U)$; **c** $X(U)$; **d** $X(U)$ and CP_Q-Autarky; **e** OLTC; **f** OLTC and CP_Q-Autarky

Control strategy	Q_t^{MV-LV} (kvar)		
	Case A	Case B	Case C
None	11.2117	24.7847	−2.5265
$\cos\varphi(P)$	11.2117	179.6204	−2.5265
$Q(U)$	26.0203	35.2988	−4.0817
$X(U)$	11.2117	24.7847	−2.5265
$X(U)+$	0.5220	15.1919	1.1476
OLTC	10.4582	24.7847	−2.0585
OLTC+	0.5354	15.1919	1.1833

Control strategy	$Q_{\Sigma,t}^{LV-CP}$ (kvar)		
	Case A	Case B	Case C
None	10.6646	9.4851	−3.6763
$\cos\varphi(P)$	10.6646	156.6212	−3.6763
$Q(U)$	25.3810	19.7681	−5.2365
$X(U)$	10.6646	9.4851	−3.6763
$X(U)+$	0.0000	0.0000	0.0000
OLTC	9.8987	9.4851	−3.2434
OLTC+	0.0000	0.0000	0.0000

Control strategy	$Q_{\Sigma,t}^{LV}$ (kvar)		
	Case A	Case B	Case C
None	0.5471	15.2995	1.1498
$\cos\varphi(P)$	0.5471	22.9992	1.1498
$Q(U)$	0.6393	15.5307	1.1547
$X(U)$	0.5471	15.2995	1.1498
$X(U)+$	0.5220	15.1919	1.1476
OLTC	0.5595	15.2995	1.1850
OLTC+	0.5354	15.1919	1.1833

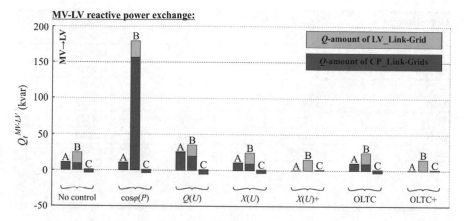

Fig. 4.130 Composition of the MV-LV reactive power exchange of the rural LV_Link-Grid for smoothed load profiles at the CP level, different cases, no control and various control strategies

Control strategy	ΔP_t^{LV} (kW)		
	Case A	Case B	Case C
None	0.7643	15.9688	1.3885
$\cos\varphi(P)$	0.7643	23.9166	1.3885
$Q(U)$	0.8266	16.4398	1.3990
$X(U)$	0.7643	15.9688	1.3885
$X(U)+$	0.7388	15.8581	1.3859
OLTC	0.7601	15.9688	1.4179
OLTC+	0.7356	15.8581	1.4159

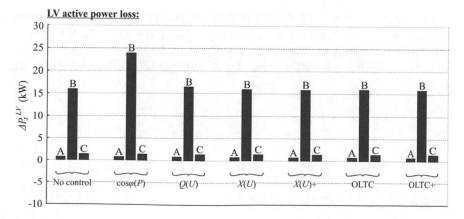

Fig. 4.131 Active power loss within the rural LV_Link-Grid for smoothed load profiles at the CP level, different cases, no control and various control strategies

Control strategy	$Loading_t^{DTR}(\%)$		
	Case A	Case B	Case C
None	13.7426	64.0142	18.4742
$cos\varphi(P)$	13.7426	77.9572	18.4742
$Q(U)$	14.8192	64.2656	18.4987
$X(U)$	13.7426	64.0142	18.4742
$X(U)+$	13.5088	63.8105	18.4553
OLTC	13.0492	64.0142	18.2095
OLTC+	12.8352	63.8105	18.1956

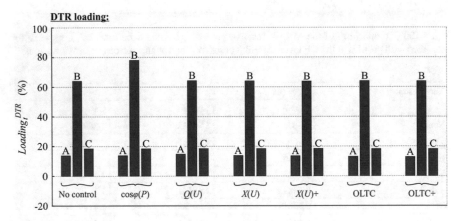

Fig. 4.132 DTR loading within the rural LV_Link-Grid for smoothed load profiles at the CP level, different cases, no control and various control strategies

Urban LV_Link-Grid

The model of the urban LV_Link-Grid is specified in Fig. 4.133; Its behaviour without any Volt/var control is depicted in Figs. 4.134 and 4.135; And its behaviour with the different Volt/var controls is shown in Figs. 4.136 to 4.138. Furthermore, Figs. 4.139 to 4.141 present the behaviour of the urban LV_Link-Grid for cases A, B, and C.

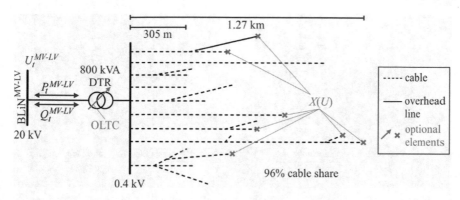

Fig. 4.133 Simplified one-line diagram of the urban LV_Link-Grid (real Austrian grid)

DTR	Rating:	800 kVA	
	Nominal voltage	Primary	20.0 kV
		Secondary	0.4 kV
	Short circuit voltage	Total	4.0%
		Resistive part	1.0%
Feeders	Nominal voltage		0.4 kV
	Number of feeders		9
	Total line length		12.815 km
	Total cable share		96.14%
	Feeder length	Maximal	1.270 km
		Minimal	0.305 km
Control parameters	OLTC	Upper voltage limit	1.025 p.u
		Lower voltage limit	0.950 p.u
		Min./mid./max. tap positions	1/3/5
		Additional voltage per tap	2.5%
	$X(U)$	Upper voltage limit	1.09 p.u
		Lower voltage limit	0.91 p.u
Connected lumped models	Link-Grids	CP	Urban residential with smoothed load profiles	175

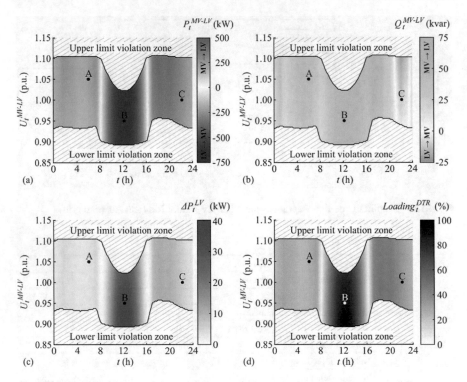

Fig. 4.134 Daily behaviour of the urban LV_Link-Grid without any Volt/var control for various voltages at the MV-LV boundary node and smoothed load profiles at the CP level: **a** MV-LV active power exchange; **b** MV-LV reactive power exchange; **c** LV active power loss; **d** DTR loading

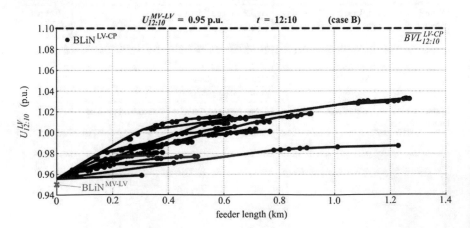

Fig. 4.135 Voltage profiles of the urban LV_Link-Grid's feeders without any Volt/var control at 12:10 for an MV-LV boundary voltage of 0.95 p.u. (case B) and smoothed load profiles at the CP level

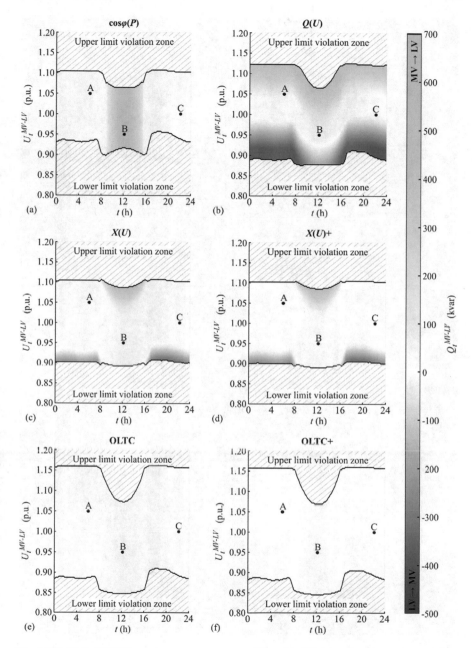

Fig. 4.136 Daily MV-LV reactive power exchange of the urban LV_Link-Grid for various voltages at the MV-LV boundary node, smoothed load profiles at the CP level, and different control strategies: **a** $\cos\varphi(P)$; **b** $Q(U)$; **c** $X(U)$; **d** $X(U)$ and CP_Q-Autarky; **e** OLTC; **f** OLTC and CP_Q-Autarky

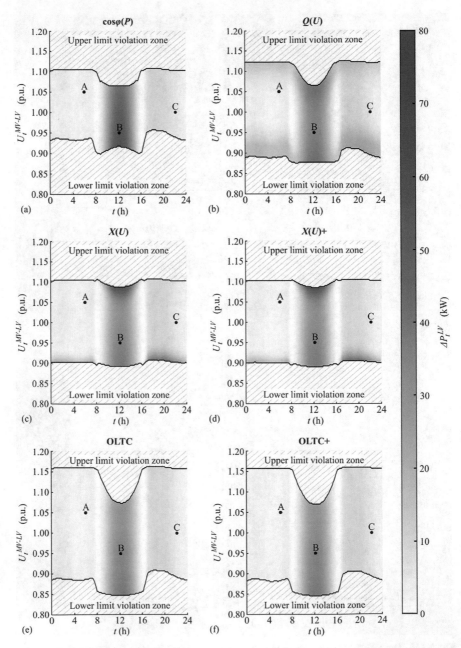

Fig. 4.137 Daily active power loss within the urban LV_Link-Grid for various voltages at the MV-LV boundary node, smoothed load profiles at the CP level, and different control strategies: **a** cosφ(P); **b** Q(U); **c** X(U); **d** X(U) and CP_Q-Autarky; **e** OLTC; **f** OLTC and CP_Q-Autarky

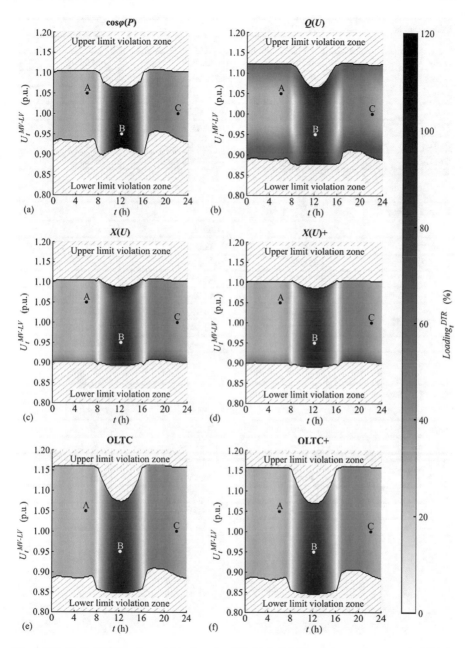

Fig. 4.138 Daily DTR loading within the urban LV_Link-Grid for various voltages at the MV-LV boundary node, smoothed load profiles at the CP level, and different control strategies: **a** cos$\varphi(P)$; **b** $Q(U)$; **c** $X(U)$; **d** $X(U)$ and CP_Q-Autarky; **e** OLTC; **f** OLTC and CP_Q-Autarky

Control strategy	Q_t^{MV-LV} (kvar)		
	Case A	Case B	Case C
None	46.5923	72.0637	−9.8095
$\cos\varphi(P)$	46.5923	512.9641	−9.8095
$Q(U)$	75.7565	66.6676	−12.6388
$X(U)$	46.5923	72.0638	−9.8093
$X(U)+$	2.7900	33.6598	5.1902
OLTC	44.9710	72.0637	−9.8095
OLTC+	2.8048	33.6598	5.1902

Control strategy	$Q_{\Sigma,t}^{LV-CP}$ (kvar)		
	Case A	Case B	Case C
None	43.6756	38.0601	−15.0114
$\cos\varphi(P)$	43.6756	459.8271	−15.0114
$Q(U)$	72.6505	32.7091	−17.8544
$X(U)$	43.6756	38.0601	−15.0114
$X(U)+$	0.0000	0.0000	0.0000
OLTC	42.0423	38.0601	−15.0114
OLTC+	0.0000	0.0000	0.0000

Control strategy	$Q_{\Sigma,t}^{LV}$ (kvar)		
	Case A	Case B	Case C
None	2.9167	34.0037	5.2021
$\cos\varphi(P)$	2.9167	53.1370	5.2021
$Q(U)$	3.1061	33.9585	5.2156
$X(U)$	2.9167	34.0037	5.2021
$X(U)+$	2.7900	33.6598	5.1902
OLTC	2.9287	34.0037	5.2021
OLTC+	2.8048	33.6598	5.1902

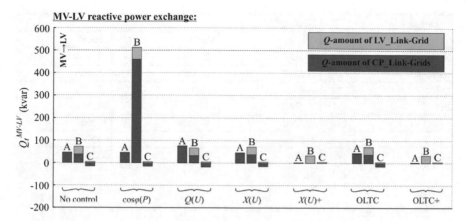

Fig. 4.139 Composition of the MV-LV reactive power exchange of the urban LV_Link-Grid for smoothed load profiles in CP level, and for different cases, no control and various control strategies

Control strategy	ΔP_t^{LV} (kW)		
	Case A	Case B	Case C
None	3.4914	34.0481	5.7219
$\cos\varphi(P)$	3.4914	53.0728	5.7219
$Q(U)$	3.6255	34.0269	5.7424
$X(U)$	3.4914	34.0481	5.7219
$X(U)+$	3.3684	33.7011	5.7095
OLTC	3.4750	34.0481	5.7219
OLTC+	3.3547	33.7011	5.7095

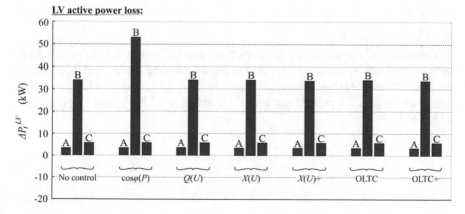

Fig. 4.140 Active power loss within the urban LV_Link-Grid for smoothed load profiles at the CP level, different cases, no control and various control strategies

Control strategy	$Loading_t^{DTR}(\%)$		
	Case A	Case B	Case C
None	27.6187	87.0541	35.3063
$\cos\varphi(P)$	27.6187	108.3297	35.3063
$Q(U)$	28.4914	86.9772	35.3293
$X(U)$	27.6188	87.0541	35.3063
$X(U)+$	27.1280	86.6339	35.2651
OLTC	26.8802	87.0541	35.3063
OLTC+	26.4107	86.6339	35.2651

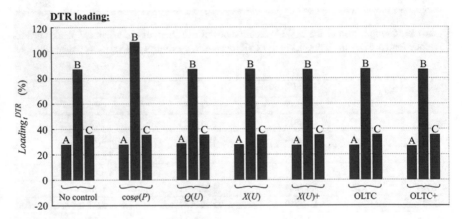

Fig. 4.141 DTR loading within the urban LV_Link-Grid for smoothed load profiles at the CP level, different cases, no control and various control strategies

MV level

Small MV_Link-Grid

The model of the small MV_Link-Grid is specified in Fig. 4.142; Its behaviour without any Volt/var control is depicted in Figs. 4.143 and 4.144; And its behaviour with the different Volt/var controls is shown in Figs. 4.145 and 4.146. Furthermore, Figs. 4.147 and 4.148 present the behaviour of the small MV_Link-Grid for cases A, B, and C.

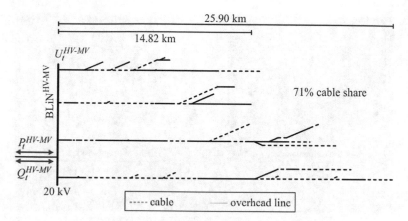

Fig. 4.142 Simplified one-line diagram of the small MV_Link-Grid (real Austrian grid)

Feeders	Nominal voltage			20 kV
	Number of feeders			4
	Total line length			158.46 km
	Total cable share			70.62%
	Feeder length	Maximal		25.90 km
		Minimal		14.82 km
Connected lumped models	Link-Grids	CP	Commercial with smoothed load profiles	90
		LV	Rural for smoothed load profiles in CP level	49
			Urban for smoothed load profiles in CP level	4

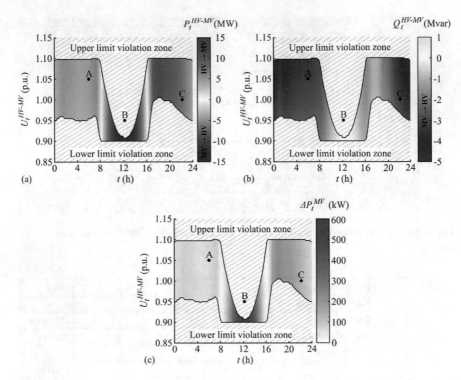

Fig. 4.143 Daily behaviour of the small MV_Link-Grid without any Volt/var control for various voltages at the HV-MV boundary node and smoothed load profiles at the CP level: **a** HV-MV active power exchange; **b** HV-MV reactive power exchange; **c** MV active power loss

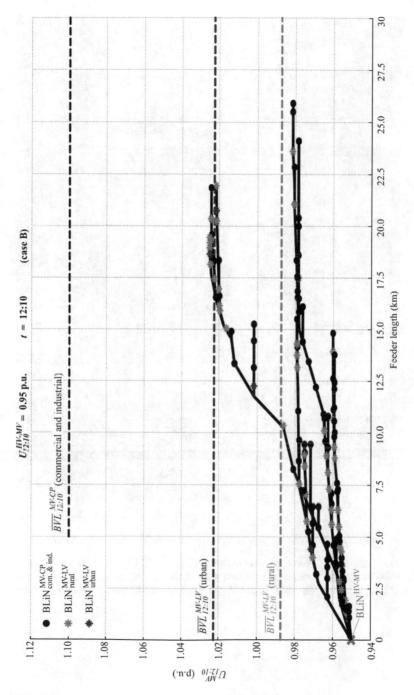

Fig. 4.144 Voltage profiles of the small MV_Link-Grid's feeders without any Volt/var control at 12:10 for an HV-MV boundary voltage of 0.95 p.u. (case B) and smoothed load profiles at the CP level

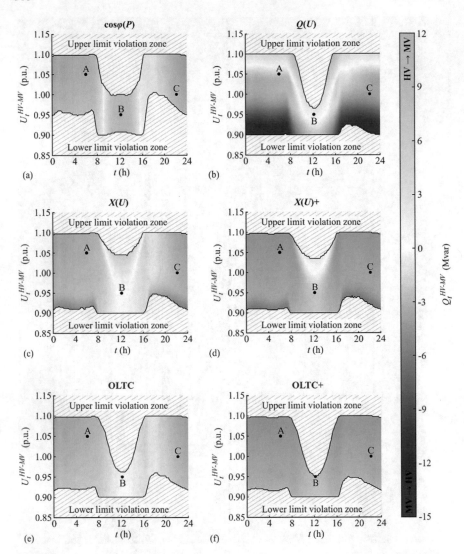

Fig. 4.145 Daily HV-MV reactive power exchange of the small MV_Link-Grid for various voltages at the HV-MV boundary node, smoothed load profiles at the CP level, and different control strategies: **a** $\cos\varphi(P)$; **b** $Q(U)$; **c** $X(U)$; **d** $X(U)$ and CP_Q-Autarky; **e** OLTC; **f** OLTC and CP_Q-Autarky

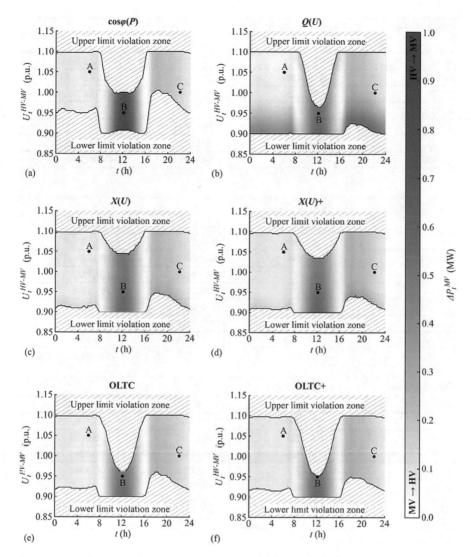

Fig. 4.146 Daily active power loss within the small MV_Link-Grid for various voltages at the HV-MV boundary node, smoothed load profiles at the CP level, and different control strategies: **a** $\cos\varphi(P)$; **b** $Q(U)$; **c** $X(U)$; **d** $X(U)$ and CP_Q-Autarky; **e** OLTC; **f** OLTC and CP_Q-Autarky

Control strategy	Q_t^{HV-MV} (kvar)		
	Case A	Case B	Case C
None	−3166.4324	−274.0876	−3158.5634
$\cos\varphi(P)$	−3166.4324	11,546.7229	−3158.5634
$Q(U)$	−1928.1602	1511.9364	−3710.0295
$X(U)$	−3166.4318	248.7131	−3158.5616
$X(U)+$	−4477.5800	−2081.4525	−3933.1527
OLTC	−3212.9555	−263.6931	−3145.0208
OLTC+	−4481.5044	−2784.5875	−3932.6640

Control strategy	$Q_{\Sigma,t}^{MV-CP}$ (kvar)		
	Case A	Case B	Case C
None	618.4366	1853.2531	999.4724
$\cos\varphi(P)$	618.4366	3974.8352	999.4724
$Q(U)$	1453.9913	1515.7483	1001.1713
$X(U)$	618.4366	1848.5563	999.4724
$X(U)+$	0.0000	0.0000	0.0000
OLTC	618.9152	1853.2941	999.5357
OLTC+	0.0000	0.0000	0.0000

Control strategy	$Q_{\Sigma,t}^{MV-LV}$ (kvar)		
	Case A	Case B	Case C
None	727.8706	1475.6636	−147.5860
$\cos\varphi(P)$	727.8706	10,826.4782	−147.5860
$Q(U)$	1127.1403	3552.6617	−702.6286
$X(U)$	727.8711	1990.3126	−147.5842
$X(U)+$	36.8124	1528.7827	78.0662
OLTC	685.6193	1484.7638	−133.5151
OLTC+	37.5357	834.2631	79.2284

Control strategy	$Q_{\Sigma,t}^{MV}$ (kvar)		
	Case A	Case B	Case C
None	−4512.7396	−3603.0043	−4010.4498
$\cos\varphi(P)$	−4512.7396	−3254.5905	−4010.4498
$Q(U)$	−4509.2919	−3556.4736	−4008.5722
$X(U)$	−4512.7396	−3590.1558	−4010.4497
$X(U)+$	−4514.3924	−3610.2352	−4011.2189
OLTC	−4517.4900	−3601.7510	−4011.0413
OLTC+	−4519.0401	−3618.8506	−4011.8923

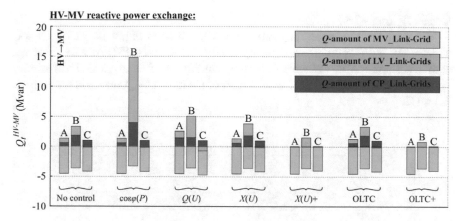

Fig. 4.147 Composition of the HV-MV reactive power exchange of the small MV_Link-Grid for smoothed load profiles at the CP level, different cases, no control and various control strategies

Control strategy	ΔP_t^{MV} (kW)		
	Case A	Case B	Case C
None	65.6263	522.4462	125.4512
$\cos\varphi(P)$	65.6263	822.7866	125.4512
$Q(U)$	56.7900	546.8396	137.9468
$X(U)$	65.6263	527.0487	125.4512
$X(U)+$	81.2532	529.9529	134.0752
OLTC	62.9133	524.2177	124.9212
OLTC+	78.1092	535.3583	133.5975

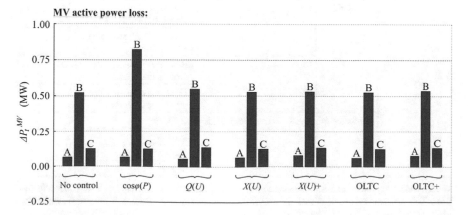

Fig. 4.148 Active power loss within the small MV_Link-Grid for smoothed load profiles at the CP level, different cases, no control and various control strategies

Large MV_Link-Grid

The model of the large MV_Link-Grid is specified in Fig. 4.149; Its behaviour without any Volt/var control is depicted in Figs. 4.150 and 4.151; And its behaviour with the different Volt/var controls is shown in Figs. 4.152 and 4.153. Furthermore, Figs. 4.154 and 4.155 present the behaviour of the large MV_Link-Grid for cases A, B, and C.

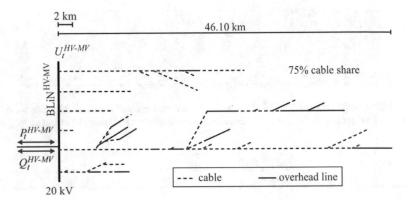

Fig. 4.149 Simplified one-line diagram of the large MV_Link-Grid (real Austrian grid)

Feeders	Nominal voltage			20 kV
	Number of feeders			6
	Total line length			267.151 km
	Total cable share			74.66%
	Feeder length		Maximal	46.10 km
			Minimal	2.00 km
Connected lumped models	Link-Grids	CP	Commercial with smoothed load profiles	143
			Industrial with smoothed load profiles	2
		LV	Rural for smoothed load profiles in CP level	45
			Urban for smoothed load profiles in CP level	11
	Producers	Hydroelectric power plants	400 kW[a]	2
			300 kW[a]	1
			100 kW[a]	11
			60 kW[a]	1

[a]The hydroelectric power plants are modelled as voltage-independent PQ node-elements. They constantly inject 70% of their maximal active power production over the entire time horizon. They do not contribute any reactive power. Their upper and lower BVLs are set to 1.1 and 0.9 p.u. for the complete simulated time horizon.

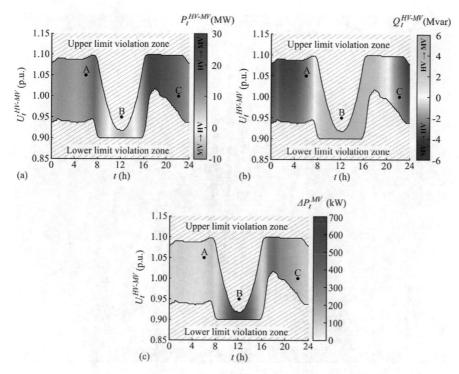

Fig. 4.150 Daily behaviour of the large MV_Link-Grid without any Volt/var control for various voltages at the HV-MV boundary node and smoothed load profiles at the CP level: **a** HV-MV active power exchange; **b** HV-MV reactive power exchange; **c** MV active power loss

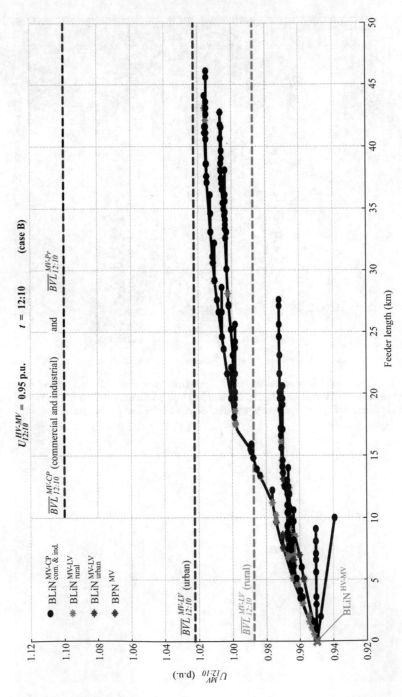

Fig. 4.151 Voltage profiles of the large MV_Link-Grid's feeders without any Volt/var control at 12:10 for an HV-MV boundary voltage of 0.95 p.u. (case B) and smoothed load profiles at the CP level

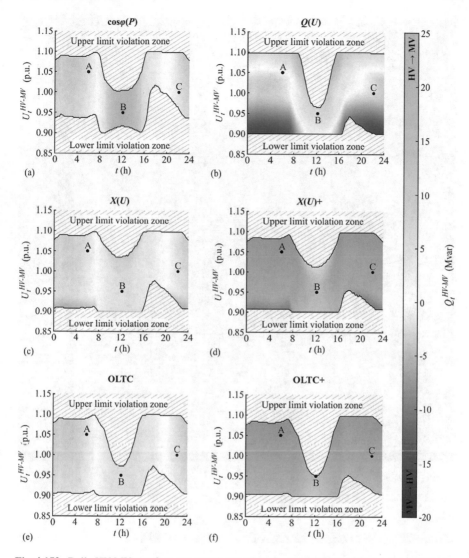

Fig. 4.152 Daily HV-MV reactive power exchange of the large MV_Link-Grid for various voltages at the HV-MV boundary node, smoothed load profiles at the CP level, and different control strategies: **a** $\cos\varphi(P)$; **b** $Q(U)$; **c** $X(U)$; **d** $X(U)$ and CP_Q-Autarky; **e** OLTC; **f** OLTC and CP_Q-Autarky

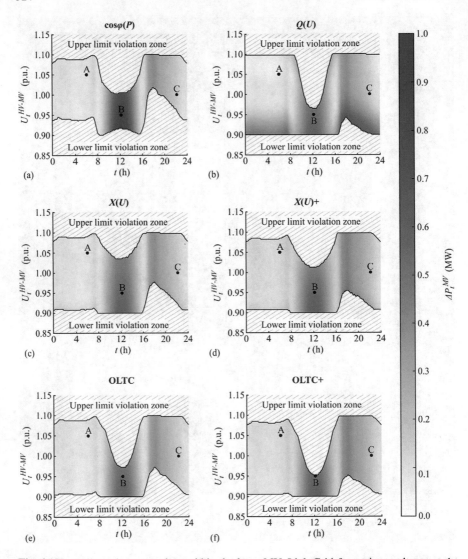

Fig. 4.153 Daily active power loss within the large MV_Link-Grid for various voltages at the HV-MV boundary node, smoothed load profiles at the CP level, and different control strategies: **a** cosφ(P); **b** Q(U); **c** X(U); **d** X(U) and CP_Q-Autarky; **e** OLTC; **f** OLTC and CP_Q-Autarky

Control strategy	Q_t^{HV-MV} (kvar)		
	Case A	Case B	Case C
None	−2178.3268	5323.0076	−2288.3236
$\cos\varphi(P)$	−2178.3268	21,186.5807	−2288.3236
$Q(U)$	112.0730	6399.8174	−2589.4373
$X(U)$	−2178.3260	5519.1924	−2288.3210
$X(U)+$	−7480.6020	−4754.2448	−6577.2863
OLTC	−2234.5103	5326.7501	−2272.0105
OLTC+	−7484.4610	−5144.4723	−6576.9824

Control strategy	$Q_{\Sigma,t}^{MV-CP}$ (kvar)		
	Case A	Case B	Case C
None	4333.0964	9521.7222	4601.8610
$\cos\varphi(P)$	4333.0964	13,099.2043	4601.8610
$Q(U)$	5899.0375	9067.4411	4604.0249
$X(U)$	4333.0964	9514.0736	4601.8610
$X(U)+$	0.0000	0.0000	0.0000
OLTC	4334.1894	9521.7074	4601.9296
OLTC+	0.0000	0.0000	0.0000

Control strategy	$Q_{\Sigma,t}^{MV-LV}$ (kvar)		
	Case A	Case B	Case C
None	1015.8510	1887.8723	−212.1237
$\cos\varphi(P)$	1015.8510	13,708.7961	−212.1237
$Q(U)$	1720.6532	3389.1906	−514.2476
$X(U)$	1015.8518	2083.9526	−212.1211
$X(U)+$	54.1633	1398.7258	109.2150
OLTC	963.1898	1891.4090	−195.2761
OLTC+	54.9485	1011.2246	110.7794

Control strategy	$Q_{\Sigma,t}^{MV}$ (kvar)		
	Case A	Case B	Case C
None	−7527.2742	−6086.5869	−6678.0609
$\cos\varphi(P)$	−7527.2742	−5621.4196	−6678.0609
$Q(U)$	−7507.6177	−6056.8143	−6679.2146
$X(U)$	−7527.2742	−6078.8338	−6678.0609
$X(U)+$	−7534.7653	−6152.9706	−6686.5013
OLTC	−7531.8895	−6086.3663	−6678.6640
OLTC+	−7539.4095	−6155.6969	−6687.7617

Control strategy	$Q_{\Sigma,t}^{MV-Pr}$ (kvar)		
	Case A	Case B	Case C
None	0.0000	0.0000	0.0000
$\cos\varphi(P)$	0.0000	0.0000	0.0000
$Q(U)$	0.0000	0.0000	0.0000
$X(U)$	0.0000	0.0000	0.0000
$X(U)+$	0.0000	0.0000	0.0000
OLTC	0.0000	0.0000	0.0000
OLTC+	0.0000	0.0000	0.0000

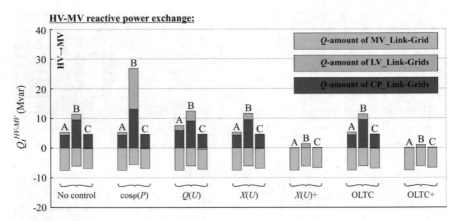

Fig. 4.154 Composition of the HV-MV reactive power exchange of the large MV_Link-Grid for smoothed load profiles at the CP level, different cases, no control and various control strategies

Control strategy	ΔP_t^{MV} (kW)		
	Case A	Case B	Case C
None	83.1072	536.9832	144.1819
$\cos\varphi(P)$	83.1072	846.0916	144.1819
$Q(U)$	52.6137	538.5407	149.2351
$X(U)$	83.1072	536.4634	144.1819
$X(U)+$	121.4523	554.7991	181.1443
OLTC	81.9192	537.2434	143.6320
OLTC+	119.6531	572.3173	180.3358

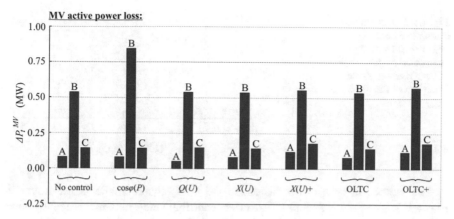

Fig. 4.155 Active power loss within the large MV_Link-Grid for smoothed load profiles at the CP level, different cases, no control and various control strategies

A.4.2 *Volt/var Control Parametrisation*

This appendix discusses the $Q(U)$ and OLTC parametrisation impact on the LV_Link-Grid's behaviour in detail.

A.4.2.1 *Impact of Q(U) Parametrisation*

The $Q(U)$-control characteristic must be adequately set to widen the BVL_t^{MV-LV} sufficiently, while avoiding unnecessary reactive power flows and oscillations. The maximal impact on the voltage is achieved when all PV inverters connected to the LV_Link-Grid contribute their maximal reactive power, either all in inductive or all in capacitive mode. However, the high reactive power flow increases the grid loss and the loading and propagates up to the superordinate MV_Link-Grid. Therefore, the characteristic should be set with a high slope gradient and a wide dead band to minimise the reactive power flows. But, too high slope gradients may lead to oscillations [39]. Figure 4.156 shows two control characteristics used to analyse the impact of $Q(U)$ parametrisation on the behaviour of the rural LV_Link-Grid: Default and customised. The default characteristic is used in Sect. 4.7 to compare the different Volt/var control strategies and yields the results shown in Fig. 4.66b. The customised characteristic is parametrised to maximise the permissible voltage range at the $BLiN^{MV-LV}$ of the rural LV_Link-Grid for the given scenario while keeping the reactive power flows as low as possible; A higher slope gradient is used in this case.

Fig. 4.156 Default and customised control characteristics used to analyse the impact of $Q(U)$ parametrisation on the behaviour of the rural LV_Link-Grid

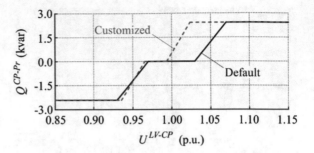

The methodology used to customise the $Q(U)$-characteristic is illustrated in Fig. 4.157. The goal is that all PV inverters contribute their maximal reactive power only when voltage limit violations occur. Simulating the scenario with smoothed load profiles defined in Sect. 4.7.2.1 for the customised characteristic yields the upper and lower BVL_t^{MV-LV} shown in Fig. 4.157a. The minimal boundary voltage that leads to upper limit violations within the LV_Link-Grid is 1.03 p.u. and appears at 12:10. Meanwhile, at 18:00 occurs the maximum boundary voltage that leads to lower limit violations, i.e. 0.9125 p.u.

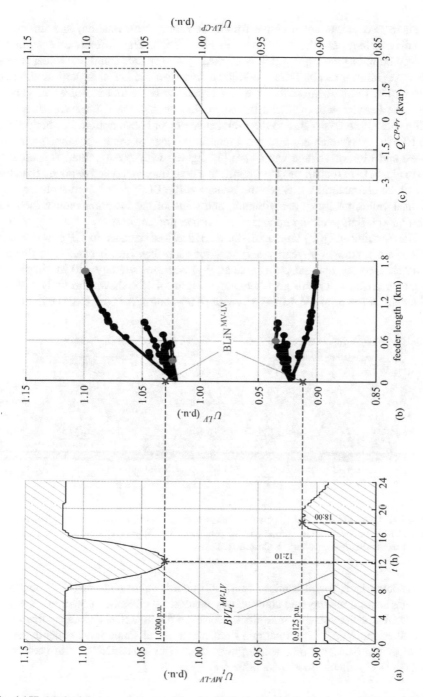

Fig. 4.157 Methodology used to customise the $Q(U)$-characteristic for the rural LV_Link-Grid: **a** BVL resulting from the customised $Q(U)$-characteristic; **b** Voltage profiles violating the lower and upper BVL; **c** Customised $Q(U)$-characteristic

These two points are decisive for the control parametrisation, and the corresponding voltage profiles are shown in Fig. 4.157b. The difference between the MV-LV boundary voltage, which is marked by a grey cross, and the voltage at the feeder beginning, results from the voltage drop over the DTR. Regarding the case with a lower limit violation, the inverters' capacitive behaviour increases the voltage along the feeders with overhead line share. In contrast, the voltage still decreases at the pure cable feeder due to the high active power consumption. For both cases, the BLiN^{LV-CP} (which are also the connection nodes of the PV systems) with the lowest and highest voltage values are highlighted with green colour. It is clear to see in Fig. 4.157c that all inverters contribute their maximal reactive power (hatched part of the characteristic). When the voltage at the BLiN^{MV-LV} comes closer to its nominal value, no limits are violated, and some of the inverters reduce their var contribution, thus avoiding unnecessary reactive power flows.

The customised $Q(U)$-characteristic is an idealised case, as the CP power contributions are unknown in reality and vary for each day. But, it enables to theoretically analyse the optimal performance of the control strategy and the impact of its parametrisation on the grid behaviour. Figure 4.158 shows the daily Volt/var behaviour of the rural LV_Link-Grid with $Q(U)$-control for both characteristics.

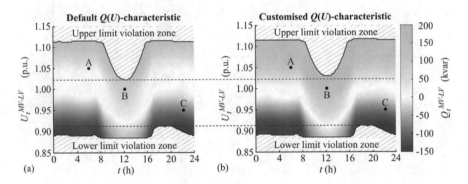

Fig. 4.158 Daily MV-LV reactive power exchange of the rural LV_Link-Grid for various voltages at the MV-LV boundary node, smoothed load profiles at the CP level, $Q(U)$-control and different control characteristics: **a** Default; **b** Customised

In comparison, the customised characteristic allows for higher MV-LV boundary voltages around midday and significantly increases the reactive power flows within a wide area of the voltage–time-plane. However, the inductive and capacitive areas, as well as the BVL_t^{MV-LV}, maintain their fundamental shape independently of the exact parametrisation. The reactive power flows over the BLiN^{MV-LV} are compared for both $Q(U)$-characteristics in Table 4.6.

Table 4.6 Reactive power flow over the BLiN^{MV-LV} of the rural LV_Link-Grid for smoothed load profiles at the CP level, different cases and distinct $Q(U)$-characteristics

$Q(U)$-characteristic	Q_t^{MV-LV} (kvar)		
	Case A	Case B	Case C
Default	26.02	78.96	−74.98
Customised	107.83	137.24	−76.15

A.4.2.2 Impact of OLTC Parametrisation

The OLTC maintains the voltage at the secondary bus bar of the DTR within a predefined voltage band, which must be adequately set to guarantee limit compliance at the LV level. The ideal parameters are found for the rural LV_Link-Grid and the defined scenario by excluding the DTR from the grid model, setting the BLiN^{MV-LV} to its secondary bus bar, and calculating the upper and lower BVL_t^{MV-LV} of the resulting model for the case without any Volt/var control, Fig. 4.159.

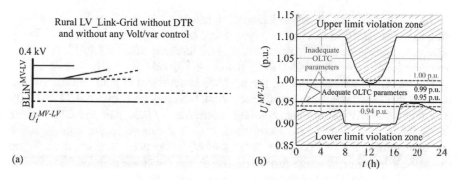

Fig. 4.159 Rural LV_Link-Grid without DTR and Volt/var control: **a** Simplified one-line diagram of the grid; **b** Daily boundary voltage limits for smoothed load profiles at the CP level

This analysis shows that no voltage limit violations occur within the LV grid when the secondary voltage stays within 0.95 and 0.99 p.u.; These values represent the adequate OLTC parameters. To study the impact of inadequate parameters, the wider voltage band between 0.94 and 1.00 p.u. is also considered in the following simulations.

The OLTC parametrisation impact is analysed by calculating the lumped model of the rural LV_Link-Grid according to Fig. 4.124 for both settings. Using the adequate OLTC parameters yields the results shown in Fig. 4.160a. The original BVL_t^{MV-LV}, i.e. those without any Volt/var control, are shifted mainly in parallel, without affecting the reactive power exchange significantly. When inadequate parameters are used, the BVL_t^{MV-LV} are also shifted in parallel, and additionally, upper and lower *limit violation islands* occur in the voltage–time-plane, Fig. 4.160b. An increase of the MV-LV boundary voltage eliminates the upper limit violations at the LV level in the upper islands. In the lower ones, a reduction of the boundary voltage eliminates the lower limit violations.

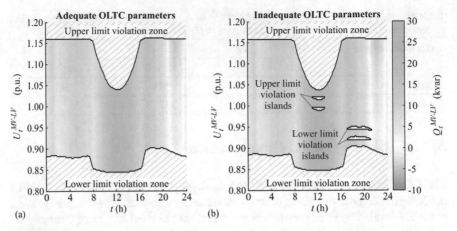

Fig. 4.160 Daily MV-LV reactive power exchange of the rural LV_Link-Grid for various voltages at the MV-LV boundary node, smoothed load profiles at the CP level, OLTC and different control settings: **a** Adequate; **b** Inadequate

To clarify the limit violation islands' occurrence, Fig. 4.161 enlarges the limit violation islands and shows the voltage profiles of all feeders of the rural LV_Link-Grid with inadequate OLTC parameters at 18:00 for three different MV-LV boundary voltages: 0.9575 p.u. (case X), 0.9475 p.u. (case Y), and 0.9375 p.u. (case Z).

In case X, the mid-position of the tap (3/5) is sufficient to maintain the voltage at the feeder beginning in the predefined band, which is set between 0.94 and 1.00 p.u. When the MV-LV boundary voltage decreases by 0.01 p.u., the CPs located at the feeder end violate their lower limit, case Y. Meanwhile, the voltage at the feeder beginning is 0.944 p.u., thus no change of the tap position is required. When the MV-LV boundary voltage further decreases, the tap changes its position to 4/5, eliminating the violations of the lower voltage limit, case Z.

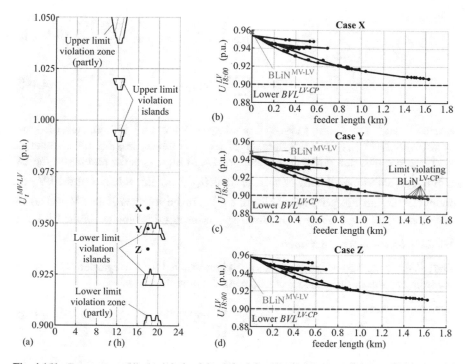

Fig. 4.161 Occurrence of limit violation islands in the rural LV_Link-Grid with inadequate OLTC parameters: **a** Enlargement of the limit violation islands; **b–d** Voltage profiles of all feeders for cases X, Y and Z

A.4.3 Volt/var control evaluation

This appendix provides the detailed definitions of the evaluation criteria used in Sect. 4.8 and the normalisation procedure used to enable their illustration within the evaluation hexagon.

A.4.3.1 Definition of Technical Evaluation Criteria

The technical criteria are calculated for the (U, t)-plane spanned by the simulated time horizon of 24 h and by the MV-LV boundary voltages between 0.9 and 1.1 p.u. (see Fig. 4.79). For brevity, the MV-LV boundary voltage (U_t^{MV-LV}) is denoted just as U in Eqs. (4.49)–(4.52).

Voltage limit violations

The voltage limit violation index (VI) is calculated for the regarded zone within the (U, t)-plane according to Eq. (4.49).

$$VI_{U,t} = \frac{1}{N^{nodes}} \left(\sum_{j=1}^{m_{U,t}} \left(\frac{U_{U,t}^{\overline{viol},j}}{U_{nom}^{LV}} - 1.1 \text{ p.u.} \right) + \sum_{j=1}^{n_{U,t}} \left(0.9 \text{ p.u.} - \frac{U_{U,t}^{\underline{viol},j}}{U_{nom}^{LV}} \right) \right) \quad (4.49a)$$

$$VI = \sum_{\forall t} \sum_{\forall U} VI_{U,t} \quad (4.49b)$$

where U and t are the MV-LV boundary voltage and the instant of time, respectively; N^{nodes} is the total number of LV grid nodes; $m_{U,t}$ is the number of the LV grid nodes that violate the upper voltage limit; $n_{U,t}$ is the number of the LV grid nodes that violate the lower voltage limit; $U_{U,t}^{\overline{viol},j}$ are the voltages of the LV grid nodes that violate the upper voltage limit; $U_{U,t}^{\underline{viol},j}$ are the voltages of the LV grid nodes that violate the lower voltage limit; And U_{nom}^{LV} is the nominal voltage of the LV level.

MV-LV reactive power exchange.

The MV-LV reactive energy exchange (E^Q) is calculated for the regarded (U, t)-plane according to Eq. (4.50) without considering the flow direction.

$$E^Q = \Delta t \cdot \sum_{\forall t} \sum_{\forall U} |Q_{U,t}^{MV-LV}| \quad (4.50)$$

where $\Delta t = 10$ min is the temporal resolution of the load profiles; And $Q_{U,t}^{MV-LV}$ is the reactive power flow through the MV-LV boundary node.

Active power loss

The energy loss (ΔE) is calculated for the regarded (U, t)-plane according to Eq. (4.51).

$$\Delta E = \Delta t \cdot \sum_{\forall t} \sum_{\forall U} \Delta P_{U,t}^{LV} \quad (4.51)$$

where $\Delta P_{U,t}^{LV}$ is the active power loss within the LV_Link-Grid.

DTR loading

The average DTR loading $(Loading^{DTR,avg})$ is calculated for the regarded (U, t)-plane according to Eq. (4.52).

$$Loading^{DTR,avg} = \frac{1}{N^t \cdot N^U} \cdot \sum_{\forall t} \sum_{\forall U} Loading_{U,t}^{DTR} \quad (4.52)$$

where N^U, N^t are the numbers of MV-LV boundary voltages and instants of time, respectively, within the regarded (U,t)-plane.

A.4.3.2 Calculation of the Evaluation Hexagon Data

The evaluation criteria defined in Sect. A.4.3.1 are calculated for each control setup (indexed with c) and both LV_Link-Grids (indexed with g) catalogued in Sect. A.4.1.2. Firstly, the evaluation hexagon is calculated for each LV_Link-Grid separately, and secondly, a common hexagon is calculated to enable the compact presentation of the final evaluation results.

Calculation of the separate hexagons
The technical evaluation criteria are normalised according to Eq. (4.53) to enable their illustration in a common chart. The resulting normalised evaluation criteria lie within the interval [0, 1] and do not have any physical unit.

$$VI_{c,g}^{norm} = \frac{VI_{c,g}}{\max\limits_{c}\left(VI_{c,g}\right)} \tag{4.53a}$$

$$E_{c,g}^{Q,norm} = \frac{E_{c,g}^{Q}}{\max\limits_{c}\left(E_{c,g}^{Q}\right)} \tag{4.53b}$$

$$\Delta E_{c,g}^{norm} = \frac{\Delta E_{c,g}}{\max\limits_{c}\left(\Delta E_{c,g}\right)} \tag{4.53c}$$

$$Loading_{c,g}^{DTR,avg,norm} = \frac{Loading_{c,g}^{DTR,avg}}{\max\limits_{c}\left(Loading_{c,g}^{DTR,avg}\right)} \tag{4.53d}$$

where $VI_{c,g}^{norm}$, $E_{c,g}^{Q,norm}$, $\Delta E_{c,g}^{norm}$, $Loading_{c,g}^{DTR,avg,norm}$ are the normalised values of the evaluation criteria of the control setup c and the LV_Link-Grid g (rural or urban).

Calculation of the common hexagon
The results of both LV_Link-Grids are superimposed according to Eq. (4.54) to enable the compact presentation of the evaluation.

$$VI_{c}^{norm} = \frac{VI_{c,rural}^{norm} + VI_{c,urban}^{norm}}{2} \tag{4.54a}$$

$$E_{c}^{Q,norm} = \frac{E_{c,rural}^{Q,norm} + E_{c,urban}^{Q,norm}}{2} \tag{4.54b}$$

$$\Delta E_{c}^{norm} = \frac{\Delta E_{c,rural}^{norm} + \Delta E_{c,urban}^{norm}}{2} \tag{4.54c}$$

$$Loading_{c}^{DTR,avg,norm} = \frac{Loading_{c,rural}^{DTR,avg,norm} + Loading_{c,urban}^{DTR,avg,norm}}{2} \tag{4.54d}$$

where VI_c^{norm}, $E_c^{Q,norm}$, ΔE_c^{norm}, $Loading_c^{DTR,avg,norm}$ are the normalised and superimposed values of the evaluation criteria for the control setup c, which are plotted in the common evaluation hexagon.

References

1. Alassi A, Bañales S, Ellabban O, Adam G, MacIver C (2019) HVDC Transmission: technology review, market trends and future outlook. Renew Sustain Energy Rev 112:530–554

2. Albarracín R, Alonso M (2013) Photovoltaic reactive power limits. In: 12th international conference on environment and electrical engineering, Wroclaw, Poland, 5–8 May, pp 13–18. https://doi.org/10.1109/EEEIC.2013.6549630

3. Alla M, Guzman A, Finney D, Fischer N (2018) Capability curve-based generator protection minimizes generator stress and maintains power system stability. In: 45th annual western protective relay conference, Spokane, WA, USA, 16–18 October, pp 1–16

4. Arif A, Wang Z, Wang J, Mather B, Bashualdo H, Zhao D (2018) Load modeling—a review. IEEE Trans Smart Grid 9(6):5986–5999. https://doi.org/10.1109/TSG.2017.2700436

5. Arrillaga J, Yonghe HL, Neville RW, Nicholas JM (2009) Self-commutated converters for high power applications. Wiley, New Jersey

6. Aunedi M, Woolf M, Strbac G, Babalola O, Clark M (2015). Characteristic demand profiles of residential and commercial EV Users and opportunities for smart charging. In: 23rd international conference on electricity distribution, Lyon, France, 15–18 June, 1088

7. Belvin RC, Short TA (2012) Voltage reduction results on a 24-kV circuit. In: IEEE PES transmission and distribution conference and exposition, Orlando, FL, USA, 7–10 May, pp 1–4. https://doi.org/10.1109/TDC.2012.6281592

8. Binder A (2012) Elektrische maschinen und antriebe: grundlagen, betriebsverhalten. Springer, Heidelberg

9. Bletterie B, Goršek A, Uljanić B, Blazic B, Woyte A, Vu Van T, Truyens F, Jahn J (2010) Enhancement of the network hosting capacity—clearing space for/with PV. In: 25th European photovoltaic solar energy conference and exhibition/5th world conference on photovoltaic energy conversion, Valencia, Spain, 6–10 September, pp 4828–4834. https://doi.org/10.4229/25thEUPVSEC2010-5AO.7.3

10. Bokhari A, Alkan A, Dogan R, Diaz-Aguiló M, de León F, Czarkowski D, Zabar Z, Birenbaum L, Noel A, Uosef RE (2014) Experimental determination of the ZIP coefficients for modern residential, commercial, and industrial loads. Trans Power Delivery 29(3):1372–1381. https://doi.org/10.1109/TPWRD.2013.2285096

11. Bollen MHJ, Sannino A (2005) Voltage control with inverter-based distributed generation. IEEE Trans Power Delivery 20(1):519–520. https://doi.org/10.1109/TPWRD.2004.834679

12. CIGRE Task Force C6.04 (2014a) Benchmark systems for network integration of renewable and distributed energy resources

13. CIGRE Working Group C4.605 (2014b) Modelling and aggregation of loads in flexible power networks

14. Carden J, Popovic D (2018) Closed-loop voltVvar optimization: addressing peak load reduction. IEEE Power Energ Mag 16(2):67–75. https://doi.org/10.1109/MPE.2017.2780962

15. Chen H, Cong TN, Yang W, Tan C, Li Y, Ding Y (2009) Progress in electrical energy storage system: a critical review. Prog Nat Sci 19(3):291–312. https://doi.org/10.1016/j.pnsc.2008.07.014

16. Choi J, Moon S (2009) The dead band control of LTC transformer at distribution substation. IEEE Trans Power Syst 24(1):319–326. https://doi.org/10.1109/TPWRS.2008.2005706

17. Choi W, Wu Y, Han D, Gorman J, Palavicino PC, Lee W, Sarlioglu B (2017) Reviews on grid-connected inverter, utility-scaled battery energy storage system, and vehicle-to-grid application - challenges and opportunities. In: 2017 IEEE transportation electrification conference and expo (ITEC), Chicago, IL, USA, 22–24 June, pp 203–210

18. Collin AJ, Tsagarakis G, Kiprakis AE, McLaughlin S (2014) Development of low-voltage load models for the residential load sector. IEEE Trans Power Syst 29(5):2180–2188. https://doi.org/10.1109/TPWRS.2014.2301949

19. Corsi S, Pozzi M, Sabelli C, Serrani A (2004) The coordinated automatic voltage control of the Italian transmission grid-part I: reasons of the choice and overview of the consolidated hierarchical system. IEEE Trans Power Syst 19(4):1723–1732. https://doi.org/10.1109/TPWRS.2004.836185

20. Dabic V, Atanackovic D (2015) Voltage VAR optimization real time closed loop deployment—BC hydro challenges and opportunities. In: 2015 IEEE PES general meeting, Denver, CO, USA, 26–30 July, pp 1–5. https://doi.org/10.1109/PESGM.2015.7286313

21. Demirok E, González PC, Frederiksen KHB, Sera D, Rodriguez P, Teodorescu R (2011) Local reactive power control methods for overvoltage prevention of distributed solar inverters in low-voltage grids. IEEE J Photovolt 1(2):174–182. https://doi.org/10.1109/JPHOTOV.2011.2174821

22. Directive (EU) 2019/944 of the European Parliament and of the Council of 5 June 2019 on common rules for the internal market for electricity and amending directive 2012/27/EU (Text with EEA relevance.). http://data.europa.eu/eli/dir/2019/944/oj. Accessed 13 Apr 2021

23. Dixon J, Moran L, Rodriguez J, Domke R (2005) Reactive power compensation technologies: state-of-the-art review. Proc IEEE 93(12):2144–2164. https://doi.org/10.1109/JPROC.2005.859937

24. ENTSO-E (2019) HVDC links in system operations. https://eepublicdownloads.entsoe.eu/clean-documents/SOC%20documents/20191203_HVDC%20links%20in%20system%20operations.pdf. Accessed 13 Apr 2021

25. EN 50160:2010—Voltage characteristics of electricity supplied by public electricity networks

26. Engelhardt S, Erlich I, Feltes C, Kretschmann J, Shewarega F (2011) Reactive power capability of wind turbines based on doubly fed induction generators. IEEE Trans Energy Convers 26(1):364–372

27. Eurelectric (2013) Power distribution in Europe: facts and figures. https://cdn.eurelectric.org/media/1835/dso_report web_final-2013-030-0764-01-e-h-D66B0486.pdf. Accessed 13 Apr 2021

28. Farivar M, Zho X, Chen L (2015) Local voltage control in distribution systems: an incremental control algorithm. In: IEEE international conference on smart grid communications, Miami, FL, USA, 2–5 November, pp 732–737. https://doi.org/10.1109/SmartGridComm.2015.7436388

29. Hingorani NG, Gyugyi L (2000) Understanding FACTS, concepts and technology of flexible AC transmission systems. IEEE Press

30. Hossain MI, Yan R, Saha TK (2016) Investigation of the interaction between step voltage regulators and large-scale photovoltaic systems regarding voltage regulation and unbalance. IET Renew Power Gener 10(3):299–309. https://doi.org/10.1049/iet-rpg.2015.0086

31. IEA—International Energy Agency (2020) Electricity generation by source, Europe 1990–2018. https://www.iea.org/data-and-statistics?country=WEOEUR&fuel=Energy%20supply&indicator=ElecGenByFuel. Accessed 13 Apr 2021

32. Ilo A (2016) Effects of the reactive power injection on the grid—the rise of the volt/var interaction Chain. Smart Grid Renew Energy 7:217–232

33. Ilo A, Schultis DL (2019) Low-voltage grid behaviour in the presence of concentrated var-sinks and var-compensated customers. Electr Power Syst Res 171:54–65. https://doi.org/10.1016/j.epsr.2019.01.031

34. Ilo A (2019) Design of the smart grid architecture according to fractal principles and the basics of corresponding market structure. Energies 12:4153

35. Ilo A, Schultis DL, Schirmer C (2018) Effectiveness of distributed vs. concentrated volt/var local control strategies in low-voltage grids. Appl Sci 8/8:1382. https://doi.org/10.3390/app8081382

36. Kundur P (1994) Power system stability and control. EPRI, McGraw-Hill, New York
37. LPF (2020) Load profile generator. https://www.loadprofilegenerator.de/. Accessed 13 Apr 2021
38. Li Q, Zhang YJ, Ji T, Lin X, Cai Z (2018) Volt/var control for power grids with connections of large-scale wind farms: a review. IEEE Access 6:26675–26692. https://doi.org/10.1109/ACC ESS.2018.2832175
39. Marggraf O, Laudahn S, Engel B, Lindner M, Aigner C, Witzmann R, Schoeneberger M, Patzack S, Vennegeerts H, Cremer M, Meyer M, Schnettler A, Berber I, Bülo T, Brantl J, Wirtz F, Frings R, Pizzutto F (2017) U-control—analysis of distributed and automated voltage control in current and future distribution grids. In: International ETG congress 2017, Bonn, Germany, 28–29 November, pp 567–572
40. McKenna E, Thomson M (2016) High-resolution stochastic integrated thermal–electrical domestic demand model. Appl Energy 165:445–461. https://doi.org/10.1016/j.apenergy.2015.12.089
41. McKenna E, Thomson M, Barton J (2015) CREST demand model. Loughborough University. https://repository.lboro.ac.uk/articles/dataset/CREST_Demand_Model_v2_0/2001129/5. Accessed 13 Apr 2021
42. Neal R (2010) The use of AMI meters and solar PV inverters in an advanced Volt/VAr control system on a distribution circuit. In: IEEE PES transmission and distribution conference and exposition, New Orleans, LA, USA, 19–22 April, pp 1–4. https://doi.org/10.1109/TDC.2010.5484402
43. Nowak S, Wang L, Metcalfe MS (2020) Two-level centralized and local voltage control in distribution systems mitigating effects of highly intermittent renewable generation. Int J Electr Power Energy Syst 119:105858. https://doi.org/10.1016/j.ijepes.2020.105858
44. Oeding D, Oswald BR (2011) Elektrische Kraftwerke und Netze. Springer, Heidelberg. https://doi.org/10.1007/978-3-642-19246-3
45. Peskin MA, Powell PW, Hall EJ (2012) Conservation voltage reduction with feedback from advanced metering infrastructure. In: IEEE PES transmission and distribution conference and exposition, Orlando, FL, USA, 7–10 May, pp 1–8. https://doi.org/10.1109/TDC.2012.6281644
46. Pflugradt N, Muntwyler U (2017) Synthesizing residential load profiles using behavior simulation. Energy Procedia 122:655–660. https://doi.org/10.1016/j.egypro.2017.07.365
47. Poliseno MC, Mastromauro RA, Liserre M (2012) Transformer-less photovoltaic (PV) inverters: a critical comparison. In: 2012 IEEE energy conversion congress and exposition, Raleigh, NC, USA, 15–20 September, pp 3438–3445
48. Preiss RF, Warnock VJ (1978) Impact of voltage reduction on energy and demand. IEEE Trans Power Apparatus Syst PAS-97/5:1665–1671. https://doi.org/10.1109/TPAS.1978.354658
49. Price WW, Chiang HD, Clark HK, Concordia C, Lee DC, Hsu JC, Ihara S, King CA, Lin CJ, Mansour Y, Srinivasan K, Taylor CW, Vaahedi E (1993) Load representation for dynamic performance analysis (of power systems). IEEE Trans Power Syst 8(2):472–482. https://doi.org/10.1109/59.260837
50. Riese P (2012) Handbuch der Blindstrom-Kompensation. https://www.frako.com/fileadmin/pdf/Downloads/Handbuch/95-00135_11_13_9066_handbuch_blk.pdf. Accessed 20 May 2021
51. Rohjans S, Dänekas C, Uslar M (2012) Requirements for smart grid ICT-architectures. In: 3rd IEEE PES innovative smart grid technologies Europe, Berlin, Germany, 14–17 October, pp 1–8. https://doi.org/10.1109/ISGTEurope.2012.6465617
52. Roytelman I, Ganesan V (2000) Coordinated local and centralized control in distribution management systems. IEEE Trans Power Delivery 15(2):718–724. https://doi.org/10.1109/61.853010
53. Roytelman I, Ganesan V (1999) Modeling of local controllers in distribution network applications. In: 21st international conference on power industry computer applications. Connecting Utilities. PICA 99. To the Millennium and Beyond (Cat. No.99CH36351), Santa Clara, CA, USA, 21 May, pp 161–166. https://doi.org/10.1109/PICA.1999.779399
54. Sarkar MNI, Meegahapola LG, Datta M (2018) Reactive power management in renewable rich power grids: a review of grid-codes, renewable generators, support devices, control strategies and optimization algorithms. IEEE Access 6:41458–41489

55. Schultis DL (2019) Daily load profiles and ZIP models of current and new residential customers. Mendeley Data. https://doi.org/10.17632/7gp7dpvw6b.1
56. Schultis DL, Ilo A (2018) TUWien_LV_TestGrids. Mendeley Data. https://doi.org/10.17632/hgh8c99tnx.1
57. Schultis DL, Ilo A (2019) Behaviour of distribution grids with the highest PV share using the Volt/Var control chain strategy. Energies 12(20):3865. https://doi.org/10.3390/en12203865
58. Schultis DL, Ilo A (2019a) Adaption of the current load model to consider residential customers having turned to LED lighting. In: 11th IEEE PES Asia-Pacific power and energy engineering conference, Macao, China, 1–4 December, pp 1–5. https://doi.org/10.1109/APPEEC45492.2019.8994535
59. Schultis DL, Ilo A (2021a) Boundary voltage limits—an instrument to increase the utilization of the existing infrastructures. In: CIRED 2021 conference, Geneva, Switzerland, 20–23 September, 910. (accepted for publication)
60. Schultis DL, Ilo A (2021b) Increasing the utilization of existing infrastructures by using the newly introduced boundary voltage limits. Energies 14(16) https://doi.org/5106-10.3390/en14165106
61. Schultis DL, Ilo A (2021c) Effect of individual Volt/var control strategies in LINK-based smart grids with a high photovoltaic Share Energies 14(18) https://doi.org/5641-10.3390/en14185641
62. Schweiger G, Eckerstorfer LV, Hafner I, Fleischhacker A, Radl J, Glock B, Wastian M, Rößler M, Lettner G, Popper N, Corcoran K (2020) Active consumer participation in smart energy systems. Energy Build 227:110359. https://doi.org/10.1016/j.enbuild.2020.110359
63. Schürhuber R (2018) Ausgewählte Aspekte des Netzanschlusses von Erzeugungsanlagen. Impulsvortrag am Treffen der CIRED, Vienna, Austria, 30 January. https://cired.at/filead min//user_upload/Ausgewaehlte_Aspekte_Netzanschluss_Impulsvortrag_Wien_Prof_Schu erhuber.pdf. Accessed 13 Apr 2021
64. Shukla A, Verma K, Kumar R (2017) Multi-stage voltage dependent load modelling of fast charging electric vehicle. In: 6th international conference on computer applications in electrical engineering—recent advances, Roorkee, India, 5–7 October, pp 86–91. https://doi.org/10.1109/CERA.2017.8343306
65. Smith JW, Sunderman W, Dugan R, Seal B (2011) Smart inverter volt/var control functions for high penetration of PV on distribution systems. In: IEEE PES power systems conference and exposition, Phoenix, AZ, USA, 20–23 March, pp 1–6. https://doi.org/10.1109/PSCE.2011.5772598
66. Sun H, Guo Q, Qi J, Ajjarapu V, Bravo R, Chow J, Li Z, Moghe R, Nasr-Azadani E, Tamrakar U, Taranto GN, Tonkoski R, Valverde G, Wu Q, Yang G (2019) Review of challenges and research opportunities for voltage control in smart grids. IEEE Trans Power Syst 34(4):2790–2801. https://doi.org/10.1109/TPWRS.2019.2897948
67. Teodorescu R, Liserre M, Rodríguez P (2007) Grid converters for photovoltaic and wind power sytems. Wiley, IEEE Press
68. Tian J, Su C, Chen Z (2013) Reactive power capability of the wind turbine with doubly fed induction generator. In: 39th annual conference of the IEEE industrial electronics society, Vienna, Austria, 10–13 November, pp 5312–5317. https://doi.org/10.1109/IECON.2013.6699999
69. Tonkoski R, Lopes LAC (2011) Impact of active power curtailment on overvoltage prevention and energy production of PV inverters connected to low voltage residential feeders. Renew Energy 36(12):3566–3574. https://doi.org/10.1016/j.renene.2011.05.031
70. Turitsyn K, Sulc P, Backhaus S, Chertkov M (2011) Options for control of reactive power by distributed photovoltaic generators. Proc IEEE 99(6):1063–1073. https://doi.org/10.1109/JPROC.2011.2116750
71. Vittal V, McCalley JD, Anderson PM, Fouad AA (2019) Power system control and stability. Wiley, New Jersey
72. Walker JH (1953) Operating characteristics of salient-pole machines. IEE Part II Power Eng 100(73):13–24. https://doi.org/10.1049/pi-2.1953.0004

73. Wang W, Lu Z (2013) Cyber security in the smart grid: survey and challenges. Comput Netw 57(5):1344–1371. https://doi.org/10.1016/j.comnet.2012.12.017
74. Wang YB, Wu CS, Liao H, Xu HH (2008) Steady-state model and power flow analysis of grid-connected photovoltaic power system. In: IEEE international conference on industrial technology, Chengdu, China, 21–24 April, pp 1–6. https://doi.org/10.1109/ICIT.2008.4608553
75. Wilson TL, Bell DG (2004) Energy conservation and demand control using distribution automation technologies. In: Rural electric power conference, Scottsdale, AZ, USA, 25 May, pp C4–1. doi:https://doi.org/10.1109/REPCON.2004.1307059
76. Xu H, Liu W, Wang L, Li M, Zhang J (2015) Optimal sizing of small hydro power plants in consideration of voltage control. In: International symposium on smart electric distribution systems and technologies, Vienna, Austria, 8–11 September, pp 165–172. https://doi.org/10.1109/SEDST.2015.7315201
77. Xu S, Wang S, Zuo G, Davidson C, Oliveira M, Memisevic R, Pilz G, Donmez B, Andersen B (2019) Application examples of STATCOM. In: Andersen B, Nilsson SL (eds) Flexible AC transmission systems. CIGRE green books. Springer, Cham. https://doi.org/10.1007/978-3-319-71926-9_13-1
78. Zeadally S, Pathan ASK, Alcaraz C, Badra M (2013) Towards privacy protection in smart grid. Wirel Pers Commun 73(1):1–22. https://doi.org/10.1007/s11277-012-0939-1
79. Zhang XP, Rehtanz C, Pal B (2006) Flexible AC transmission systems: modelling and control. Springer, Heidelberg
80. Zhang F, Guo X, Chang X, Fan G, Chen L, Wang Q, Tang Y, Dai J (2017) The reactive power voltage control strategy of PV systems in low-voltage string lines. In: IEEE manchester powertech, Manchester, UK, 18–22 June 2017, pp 1–6. https://doi.org/10.1109/PTC.2017.7980995
81. Zhou X, Farivar M, Liu Z, Chen L, Low SH (2021) Reverse and forward engineering of local voltage control in distribution networks. IEEE Trans Autom Control 66(3):1116–1128. https://doi.org/10.1109/TAC.2020.2994184

Correction to: Introduction

Correction to:
Chapter 1 in: A. Ilo and D.-L. Schultis, *A Holistic Solution*
for Smart Grids based on LINK–Paradigm, **Power Systems,**
https://doi.org/10.1007/978-3-030-81530-1_1

In the original version of the book, the following belated correction has been incorporated: In Chapter 1, Figure 1.7 has been replaced with a new image. The book and the chapter have been updated with the change.

The updated version of this chapter can be found at
https://doi.org/10.1007/978-3-030-81530-1_1

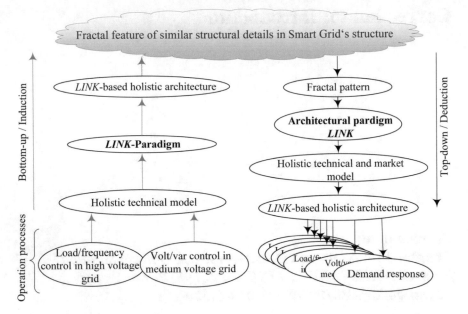

Fig. 1.7 Overview of the bottom-up and top-down methods used to find the *LINK*-Paradigm and design the corresponding holistic architecture

Glossary

Automatic is having the capability of starting, operating, moving, etc., independently.

Automation is the technique, method, or system of operating or controlling a process by highly automatic means, as by electronic devices, reducing intervention to a minimum. (British Dictionary; https://www.dictionary.com/browse/automation?s=t)

Complete Smart Grid solution is a solution that guarantees a stable, reliable and cost-effective operation of a more environmentally-friendly smart power system. It should also have the ability to ride through the transition phase and further on without causing any problems.

Control Local Control refers to control actions that are done locally (device level) in an open loop. The control is characterised by the open action path, in which the output variables influenced by the input variables do not act on themselves continuously and again via the same input variables. This kind of control is often used to control reactive devices, (e.g. switched capacitors banks, where the output variable is always reactive power, while the input variable may be voltage, current, time and so on).

Primary Control refers to control actions that are done locally (device level) in a closed loop. The control variable is locally measured and continuously compared with the reference variable, the set-point. The deviation from the set-point results in a signal that influences the valves, excitation current, transformer steps, etc. in a primary-controlled power plant, transformer, and so on, such that the desired power is delivered or the desired voltage is reached. The characteristic of the primary control is the closed action sequence, in which the controlled variable continuously influences itself in the action path of the control loop.

Secondary Control refers to control actions that are calculated based on a control area. It fulfils a predefined objective function by respecting the constraints of electrical appliances (PQ diagrams of generators, transformer rating, etc.). Secondary control calculates and sends the set-points to primary controls and the input variables to local controls acting on its own area.

© Springer Nature Switzerland AG 2022
A. Ilo and D.-L. Schultis, *A Holistic Solution for Smart Grids based on LINK–Paradigm*, Power Systems,
https://doi.org/10.1007/978-3-030-81530-1

Digitisation is the process of transcribing data into a digital form (0, 1) so that it can be directly processed by a computer.

Digitalisation is the process of moving to a digital business that is using digital technologies to change business models and provide new revenue streams and value producing opportunities.

Paradigm is a symbolic model or diagram that makes it easier for us to understand the essential characteristics of a complicated process.

Smart City is a developed urban area that creates sustainable economic development and high quality of life by excelling in multiple key areas; economy, mobility, environment, people, living, and government.

Distribution transformer is a transformer connecting medium and low voltage grids.

Supplying transformer is a transformer connecting high and medium voltage grids.

Consuming device is a device converting electricity into a service for the end-user.

Index

© Springer Nature Switzerland AG 2022
A. Ilo and D.-L. Schultis, *A Holistic Solution for Smart Grids based on LINK–Paradigm*, Power Systems,
https://doi.org/10.1007/978-3-030-81530-1

Printed in the United States
by Baker & Taylor Publisher Services